Understanding Marketing & Technology Without Losing Your Mind

A Practical Guide for Managers and Leaders

Dave Tedlock

Printed in the United States of America

First Printing, April 2013

ISBN 978-0615762616

San Marcos Press
3138 North Swan Road
Tucson, AZ 85712

www.sanmarcospress.com

Understanding
Marketing
&
Technology
Without Losing Your Mind

A Practical Guide
for Managers
and Leaders

Dave Tedlock

For Sandy and Michael

Understanding Marketing & Technology Without Losing Your Mind
A Practical Guide for Managers and Leaders

Contents

Preface and Acknowledgements

This Book Got Started Without Anyone Noticing

Years ago, when I first thought of writing a column about the Internet and marketing, I contacted Eric Snyder, the news editor of *Inside Tucson Business*. My goal was to share insights I'd gained by working through challenges with my own clients and by helping colleagues.

Eric liked my sample column well enough to start publishing the column twice a month. I didn't know it then, but the book you're reading now got its start with that first effort.

A Wide Range of People Liked It

In just weeks, we learned that all sorts of people liked the column. Marketing, advertising, communication and public relations people, working within organizations or providing organizations with services, reacted positively. They called, emailed or just stopped me on the street and expressed appreciation.

Owners and Managers in Small Organizations Liked It

Small-business owners and managers, and their counterparts in nonprofit organizations, reacted the same way. Sometimes a thank you came a year or even two after a column had appeared.

Even IT People Liked the Columns

The surprise for me was that Information Technology (IT) people also became readers. I had worried they might find the information simplistic, the insights uninspired. Instead, some were just grateful. IT people are responsible for such a huge array of technologies. Keeping up with all those technologies can be extremely challenging.

When I decided to offer the column to business publications outside of Tucson, several people volunteered to write a comment about the column or agreed I could use a comment they had already emailed me. These professionals ranged from, for example, Steve Orton, a VP of sales and marketing at The River, to Erin Fitzgerald, a marketing director. A marcom professional, Roger Yohem, wrote about the Cousin Billy column, "It's hilarious. It brought tears to my eyes, made my belly hurt and blew my concentration for the whole rest of the day!"

The column appeared for a time in *The New Mexico Business Weekly*, *Idaho Business Review* and occasionally in *The Arizona Business Gazette* (in Phoenix).

I Took Breaks and My Editors Changed

After taking a break from writing the column, I began again. This time, Teresa Truelsen, an editor at the *Tucson Citizen*, asked me to write for a monthly magazine, *Tucson Business Edge*. After yet another break, I came full circle as a columnist. The column began to appear, once again, in *Inside Tucson Business*, at the invitation of its editor, David Hatfield.

A Marcom Man Pushed for a Book

Then one day Roger Yohem urged me to review all the columns I'd ever written, select the best and make them into a book. He was enthusiastic and persuasive. Thanks, Roger, for pushing me. At about that time, Cecilia Vindiola, who was retiring as a community relations outreach coordinator, emailed me, "I so love the work that you do and I don't want to miss a word! You totally rock!"

Many People Helped

Countless other people have been supportive. Some stand out, especially Jodi Horton, APR, a Fellow at PRSA and president of her own company, Ideas @ Work Inc. Jodi seems enthusiastic about nearly every idea I share with her, but even taking that into account, she seemed genuinely thrilled by prospect of this book. Wendy Erica Werden, whose marketing positions include being the director of marketing and strategic partnership at Arizona Public Media, also provided encouragement.

Jon Black, my vice president at NetOutcomes, quickly became indispensable during the production of this book. He has served as a researcher, technical consultant, editor, book designer and idea man. Jon put his remarkable range of talents to good use on this book.

I've also been blessed to have Ford Burkhart provide copyediting and great advice about many key (sorry Ford) issues.

Michael Kalinowski, Kalinowski Design, provided design work for the cover and laughed (in the right way) at the title. Claudia Velasquez was a capable proofreader. Additional technical input and opinion came from two of the brightest and most capable technical guys I've ever met, Ed Schaeffer, the owner of Better Bytes, and John Moffatt, computer consultant.

Get the Latest

Even with all this help, there are errors in this book. The mistakes belong to me. The ideas belong to anyone who reads the book. Thank you, in advance, for becoming a reader. To read my latest column or sign up for our free newsletter, visit my blog at *www.netoutcomes.com*.

The Target Market for this Book, Reading Tips and More

This book can be valuable to you if:

1. You work in an organization as a communication, public relations, advertising or marketing person.
2. You work in a public relations firm, advertising agency or digital marketing company.
3. You are a leader in a small business or nonprofit and your duties include managing marketing, communication and/or technology processes and decisions.

You may get special value from this book if your organization is too small to have one or more full-time advertising, marketing or public relations people on staff, or too small to have a full-time technology person.

You should find this book especially useful in helping you understand the activities, conflicts, pitfalls and challenges that occur at the intersection of technology and marketing.

That intersection — the crossroads of technology and marketing — is a dangerous one. It can be hazardous to your organization and to your career.

You deserve to have these complex topics explained in an engaging way. You also deserve to feel a little entertained along the way. Fun helps.

One of the readers of my column told me, "I want you to write about what those IT [Information Technology] people aren't telling me. What is it they don't want me to know? What secrets do I need to know?"

This book tries to explain what usually is not explained well and reveal secrets that shouldn't be kept.

Another reader said, "You may not explain every detail, but I know what questions to ask."

If you are in IT, you may find value here if the topics are new to you. If all of these topics are familiar, you may find yourself disagreeing with me from time to time, or perhaps all of the time.

IT people often don't understand what nontechnical people need to know, how much they need to know and why they need to know it. Unfortunately, some IT people *like* to keep secrets. How is that good for the organization?

Tips on Getting the Most Out of This Book

1.   The table of contents organizes these pieces by topic, but each piece stands on its own. Review the topics and read the piece important to you right now. Keep the book on hand because a different topic will

become important later.

2. The topics are arranged in alphabetical order. Start wherever you like.

3. To get the new Dave Tedlock blog posts, subscribe to my newsletter at *www.netoutcomes.com*. It's free. You can unsubscribe whenever you like.

The Story behind This Book

For several years now, with a few breaks, I've been paid to write a column for one or more business newspapers, including Tucson Business Edge, the New Mexico Business Weekly, the Idaho Business Journal and Inside Tucson Business. I kept writing mostly because of all the positive feedback I got from readers.

One day not too long ago, a friend of mine, a journalist, urged me to pull the best pieces together and publish a book. He deserves a lot of thanks, so here's one: Thanks, Roger Yohem. You rock.

From more than 100 columns, I selected the best, the ones that continue to be strong and relevant. Then I revised them all. Finally, I wrote several new pieces to explain topics that seem especially important.

Even so, you may not find the topic you want to know about covered here. This is a book of selected writings, not an attempt to recreate Wikipedia. If you want me to address a topic, email me at *dave.tedlock@netoutcomes.com*. I'll see what I can do.

Lastly, thanks for reading this far, whoever you are. Now keep going!

Chapter 1

Advertising and Marketing Online

Would Wanamaker Love Internet Advertising?

Before the Internet, the CEO of one of Arizona's largest credit unions called me about the ad campaign Tedlock Advertising, my agency, had recommended to promote the credit union's used car loan rates.

"I looked at your creative and production costs," he said. "Instead of paying you for that, can we just pay you for the number of car loans we actually write?"

Try thinking through the issues that arise: How many car loan prospects exist in the market place? Are this organization's rates competitive? How many prospects would be eligible to join the credit union? What percent of those people would qualify for a loan? What would their member experience be like? How capable are the loan officers at closing? How would actual loans written be counted? Would other products sold (cross-selling opportunities have value) also be measured? What about the dollar value of this car loan campaign in terms of branding and product positioning?

We talked. I may have quoted the statement made by department store merchant John Wanamaker, "Half the money I spend on advertising is wasted; the trouble is I don't know which half." It so happens my client cancelled the campaign because my agency wasn't willing to bill him based on loans written.

After Google and Facebook, organizations can measure their Internet marketing with great precision compared to what Wanamaker experienced. We've enabled one client to measure, for example, the exact number of new customers generated directly from a click-through on a banner ad resulting in a scheduled appointment.

It is tempting to point and say, "Wow, precision like that makes online marketing highly effective. Wanamaker would love it." Reaching that kind of conclusion, however, may involve a false leap in logic. Just consider some of the variables.

One, a click-through gets people to a website, but the website still must deliver the design, organization, content and programming necessary to generate and accept the sale. In other words, the "visitor experience" must attract and keep people. Here's an example. One way content can sell is through service or product reviews. Research shows that our buying decisions are influenced strongly by reviews even from people we don't know.

We once lost a large project because we refused to drive traffic to a weak website, a site that was an entire generation behind its competition. We strongly recommended upgrading the site first and then driving traffic to it with the balance of the budget. Instead, our prospective client decided to spend 100 percent of the money driving traffic to a bad site. Six months later, the prospective client glumly reported that the ad campaign was a complete flop — no leads were generated.

Then, too, a sale can only take place when the visitors who click through are qualified to buy the product. Not long ago my son spent considerable time on at *Lexus. com* looking at the LFA, but my first comment when he insisted I watch the video was this: "That car's priced at $375,000!" So my son made a wonderful contribution toward average time on site but he was not a prospective buyer, nor is he likely (I'm an optimist, honest!) to ever become a prospect.

Even when qualified visitors arrive, we may lose the sale and not know why. Sure, at times we get lucky and know. For example, an e-commerce site may reveal that the largest number of abandoned shopping carts come from the "shipping costs" page upon checkout. Follow-up research may reveal that competitors offer a similar product with free shipping. A change of strategy may be required.

Typically, however, our wide and diverse array of measuring tools produces a murkier picture. We launch a complex Search Engine Optimization (SEO) campaign, but have to wait months to see results. Then suddenly the client tells us online sales have increased to 27 times the original amount. Is this huge increase from the site or other factors?

Thus many factors come into play: product or service reputation, name recognition built up by both on-line and off-line marketing, competitive price point and the ability of the organization itself to actually close a sale when a buyer appears, in person or online.

As I wrote this column, a client called to ask about an online medium wanting to renew its banner advertising agreement. I asked the client: What are the click-through rates? Are click-throughs tied to lead generation? What are the conversion rates to sales? Or, if any of those are not being measured now, how can they be? In sum, organizations can and should ask their Internet marketing firms to measure and report in detail on results. Measuring and reporting results is a worthwhile investment.

Before the Internet, Wanamaker felt he could waste half his money because the other half worked. Today, no organization can afford to be that sloppy. The new challenge is to choose wisely from an ever-widening array of options and measure the results of that investment. We have to work like crazy not just to measure twice, but maybe to measure 16 times, to cut once — right to a sale.

Facebook Surprises Us, and Google

Read the following items about Facebook and decide which one surprises you the most. Think of it as a poll. Even if your answer is "none," read on for other surprises about Facebook and its battle against Google for Internet dominance. What surprises you the most about Facebook?

1. It's been valued at $50 Billion or more.
2. In 2010 its ad revenue, according to eMarketer, was $1.86 Billion and has risen sharply since.
3. People who fill their Facebook account with tremendous detail about themselves are offended when Facebook uses the data to target advertising at them or use them for testimonials.
4. Facebook's page views and unique visitors rival Google's.
5. Facebook's ads wear out five times faster than Google ads.

Few of us are financial analysts, so let's skip over Facebook's current value. We can be a little cautious about Facebook's ad revenue figures because reports are new, but certainly the numbers blow away any lingering doubt about whether Facebook has a revenue model.

Now consider the fact that hundreds of millions of people have filled their Facebook pages with all kinds of details about themselves: their photos, education, marital status, employment status, job title, birth date, hobbies, interests, home town, likes and dislikes. Facebook offers advertisers the opportunity to selectively advertise to people based on these highly detailed profiles.

Given the billions Google generates in ad revenue, it's worth comparing Google advertising to Facebook's. Facebook advertising is different. As early as September 2010 in Entrepreneur magazine, Elsa Wenzel pointed out the fundamental difference. She writes, "Google AdWords ... matches keywords on the pages of Google search results, whereas Facebook Ads can match specifics in a user profile." Put simply, Google ads target people who are searching for something. Facebook ads can target people in a very specific demographic.

According to WebTrends research, ads on Facebook wear out five times faster than ads on Google. That makes sense. Facebook ads tend to stagnate on pages whereas Google ads get shuffled based on searches.

Now compare Facebook to Google in two other key ways. According to a Hitwise report, Facebook produces more page views than Google. That's huge. Another report, from comScore back in October 2010 shows Facebook rivaling Google for unique visitors, with 151.13 million U.S. uniques for Facebook, 173.3 million uniques for Google.

For years marketers wondered whether there would ever be an Internet "Pepsi" to rival Google as "Coke." The answer is YES — Facebook is Pepsi to Google's Coke in terms of Internet traffic, but Facebook is, of course, a different product and opportunity.

Meanwhile, Facebook grows more aggressive by the day. Now advertisers can use the Facebook feature "like" (Facebook killed "Become a Fan"). The new ad product is called Sponsored Stories. Here's a simplified explanation of how it works. Suppose Nothing-But-T Shirts wants to advertise on Facebook and has worked diligently for the past year acquiring 50,000 people who "like" Nothing-But.

Now, Nothing-But can run an ad campaign in which many of those 50,000 Facebook users may involuntarily become spokespersons for Nothing-But's latest product.

Where does the ad run? On the pages of the Friends of those 50,000 people. So if Bob Smith has 257 friends, all 257 could see an ad on their page in which Bob is promoting Nothing-But-T Shirts. Bob doesn't have a choice — this is not a privacy settings issue. For Nothing-But though, Sponsored Stories could be a great marketing tool. Research shows we value most recommendations that come from people we know.

Suppose the average person has 200 friends, and

Nothing-But can buy into 40,000 of those people's friends. Nothing-But can then run testimonial ads on 8 million pages. Google can't offer an advertising package like that. Of course, millions of Facebook users are upset. Some may feel there's a difference between "liking" something and agreeing to become a spokesperson for the service or product for no pay. Some say they don't like the idea of annoying their friends by having an ad, with their name on it, appearing on their friends' pages.

What we now know, and should have anticipated, is that Facebook and Google, and the people who use them, will continue to surprise us. In the future, the only surprise for us will be if there are no more surprises.

Free Advertising on Google, Yahoo! and Bing Is Neglected, Even Wasted

Online, the most valuable advertising space that exists today may be the first page of organic search results for Google (65 percent of searches), Yahoo! (13+ percent of searches) and Bing (13+ percent of searches). Billions of hours – and dollars – get spent every year in a never-ending effort to have a particular organization's listing come up on Page 1 of search results when a person types in, say, "hotels Tucson."

It's easy to understand the geography of search engine results. A typical results page has three sections: paid advertisements (Sponsored Links), directory listings (Google maps, Yahoo! and Bing Local) and organic results. "Organic results" is geek-speak for the "free" listings appearing below sponsored links and (above or below) directory listings.

Organic results are truly free advertising in the sense that the search engine charges nothing to the organizations listed there. What's surprising is that some organizations do not take full advantage of these free "ad listings." Regardless of whether the result appears at the top of Page 1 or the bottom of Page 7, FREE is still a big deal.

An organic listing has two key parts. One is what's called the "title." It should be written as an advertising headline. Typically, it appears in bigger, blue type in search results. At the time I did a sample search for "hotels Tucson," for example, a listing appeared at the bottom of the Page 1 with the ad headline "The Historic Hotel Congress" in blue text.

The second part of an organic listing is the "description," which can be up to two lines of black text. In ad terms, this is really short-form sales copy. In this "hotels Tucson," example, the description given at that moment in time was "Renovated hotel built in 1919, with rooms decorated with vintage items. Photographs and details of rooms, restaurant, dance club and banquet facilities."

There's a third part, of course, the link to the site, but nearly everyone gets that right, so let's focus on the ad headline ("title") and sale copy ("description"). Search engines get the text for this "title" not from the words people see when they actually read a website page, but from text placed in the code at the top of that page. Search engine bots dutifully read, record and report this information under the label "page title."

In the Hotel Congress organic listing a small failing is that these headlines (page titles) can be up to 60 characters long, yet Hotel Congress only opts for 27. Count 'em: "The Historic Hotel Congress." The ad is free, but Hotel Congress uses less than half of its free space. That's like having a 14 x 48 foot billboard (a standard size) and putting up a design that uses only half the space. Why not us it all? For example, "Historic Hotel Congress. Tucson. Reservations: 800.722.8848" would fit. Which one sells better?

Now consider the description. Search engines generally pull the description from the page code (source) as well, in a section labeled "description." In our example, the organic result dutifully reported is, "Renovated hotel built in 1919, with rooms decorated with vintage items. Photographs and details of rooms, restaurant, dance club, and banquet facilities."

Uh, is "yuck" too strong a reaction?

Oddly enough, the home page itself (at one time) has this lead copy for humans to read, "Located in the heart of sunny, downtown Tucson, Arizona, where summer spends the winter, the Hotel Congress is at the hub of Tucson history and nightlife."

Now, you decide. Which copy makes you want to go there? I vote for the place that's summer in winter at the hub of history and nightlife. In short, Hotel Congress needs to get this sentence inserted into the code and, typically, the search engines will dutifully position the hotel much more effectively.

The point: on every page, within reason given page content, organizations can write a free organic search result ad for themselves in the page title and description.

Yet often the space is neglected or misused.

Now, go check out your own site. Google the name of your business, or better yet use your domain name, and look at the organic listings that appear. Do your page titles use the full 60 characters? Does your description use 140? Do the titles/descriptions advertise/sell what's on that page? Free advertising is hard to come by. Make the most of yours, page by page. It seems silly not to, doesn't it?

Here's a postscript on Hotel Congress. Years later, even after actually emailing me a thank you, the latest Hotel Congress ad headline (page title) is, "Hotel Congress, Tucson, AZ." Well, at least the word "Tucson" is in there. And the description? Well, nothing's changed. The body copy still says, "Renovated hotel built in 1919, with rooms decorated with vintage items. Photographs and details of rooms, restaurant, dance club, and banquet facilities." The slide show on the home page positions the hotel in a completely different, attention-getting way. The problem is that the organic search result might not inspire people to go there.

Chapter 2

Blogs

Making Blogging Picture Perfect: Putting the Right Pieces in Place

As early as 2009, some online marketers suggested that the power of social media had rendered blogs mainstream — an old approach and ineffective. That's nonsense.

Executives and marketing/communication people should consider blogging. Blogs can and do have a huge impact on search engine ranking, customer retention and sales. Blogs can be so powerful and so effective, in fact, that you should be asking, "So, why doesn't everyone blog and succeed?"

The bloggers who fail — and most do — don't take the time to plan and then execute as needed. Consider this case study about XYZ Corporation (XYZ) complete with a surprise, at the end, about XYZ's type of business and its target market. Note how each piece of the blog puzzle must be in place for a picture-perfect, successful blog to emerge.

One piece of this puzzle is leverage. We usually define a blog as a series of written "posts" placed on a website. XYZ leverages its blog by simultaneously making it an email campaign that reaches its core target market.

XYZ maintains the blog on its website, which is a passive marketing tool aimed at humans and bots combing through it for search engines. XYZ's email campaign (opt-in/opt-out of course) is not redundant because email is proactive. Email campaigns also, by the way, have much higher closing ratios than do social media efforts and other marketing tools.

The second key factor is frequency. Many experts recommend blogs be completed at least weekly. One extremely successful blogger we know of recommends blogging every five days, about 70 posts a year. XYZ posts a new blogs 52 times a year. That high level of frequency is a lot of work.

Regardless of whether you blog 52 times a year or 70, you can see that "commitment" is another key to success. That's why more than 90 percent of all blogs that get started fail later on. The work load is high. One way the XYZ Corporation copes is to have the main blogger take a break once a month, with an associate blogger filling in.

You can see that making a blog successful is not a sprint. It's not even a marathon. Writing a blog is more like a participating in a series of marathons, or walking from coast to coast in a year. Week in, week out, month in, month out, over and over again, the post must get written. The XYZ Corporation was blessed to have a blogger committed enough to stay the course, and enough talent to write a good post.

Talent is a fourth vital element. Each post is, fundamentally, a substantial writing assignment. An ideal blog should be between 400 and 700 words per post. The XYZ Corporation maintains its quality by writing an even shorter blog of just 100 words. It works.

If a blog is badly written, bots will still read it, but people won't. Today more than ever, we all struggle to cope with a sea of content that surges around us. One way we cope is to avoid bad content. When we find great content, we pay great attention.

Fifth, content matters. In short, the content of the blog must be important and valuable to the intended audience — to the readers of the blog. This requirement means, therefore, that the blog's over-arching topic must be broad enough to offer topics for consideration up to 70 times a year. That's 70 ways to talk about "x."

Before you begin a blog, ask yourself whether you can think of 50 topics you might write about. Suppose you are the marketing and sales director for a retailer selling car and truck tires. Do you think you could come up with 50 posts about tires? Most of us don't buy tires often enough to care all that much about reading about them. If we don't care, we don't read. XYZ Corporation, on the other hand, has an almost limitless supply of topics to blog about, topics that speak directly to its target market.

Once you have a successful blog, there are a dozen ways to leverage that content: Tweets, e-books, white papers, Facebook posts and more. These are additional payoffs, but none of these matter if the five fundamental requirements haven't been met first.

Now, here's the revelation (pun intended) about the XYZ Corporation. XYZ is actually Tanque Verde Lutheran Church in Tucson, Ariz. The blog/email campaign is called Grace Notes. Grace Notes is posted on a website, *tvlc.org*, and emailed to subscribers (leverage). It's written weekly by Pastor Gayle Bintliff (a talented writer) with great commitment. Greg Mannel, the associate pastor, steps in once a month to give her a break.

The target audience is the membership of the church. Each topic is a preview of the week's upcoming sermon. There are plenty of topics to cover, as you can imagine, given that the sermons are based on a whole book full of content (the Holy Bible).

Just try Googling "Pastor Gayle, Grace Notes" and you'll see the search engine ranking is high. But you don't need a sermon on blogging — you just got one.

Chapter 3

Cloud Computing

What the Cloud Is and Why You Should Care

You go on a working vacation, take your laptop to the beach to knock off a few projects and are surprised to discover your escape from work is so complete you have no Internet access. There's no Wi-Fi on the beach and your smartphone connection is too weak to provide a hotspot. You can't receive new email or even access the thousands of emails, and their attachments, you've carefully saved for years. As a result, you can't get any real work done. You'll have to settle for getting some sleep and a tan. Welcome to the Cloud!

Your assistant makes a disturbing confession. At your direction, she deleted your (former) communication coordinator's WordPress account last week -- you didn't want him to be able to access the company's website. The problem is, your assistant failed to assign his pages and posts to another account. The result is that the last nine weeks of his website edits are gone! Welcome to the Cloud!

You get an emergency call from your security service telling you there's been a break-in at your office. Despite your security company's guard arriving at your place seven minutes later, thieves had time to grab your network file server AND your backup drives. For some organizations, this double theft could mean that vital company files are gone. You, however, make sure a separate backup runs over the Internet to a server in another city. It'll take some work but you can restore every single file you need. Welcome to the Cloud!

These three examples — about email, websites and network backup files — provide three ways to understand what the Cloud is and why you should care. Not everyone agrees upon how to define "the Cloud," but these examples give you a way to understand it.

First, consider email. Simply put, typically email is delivered to your computer, tablet or smart phone in one of two ways: POP (Post Office Protocol) or IMAP (Internet Message Access Protocol). If you have POP email (usually called POP3), new incoming email waits for you on a server, but when you download it, your email lives on your computer and/or smartphone. Thus you can easily save email on your computer and access it regardless of whether you have Internet access.

IMAP email often works differently (though the description here is simplified). IMAP gives you access to your email files from your smartphone / laptop / desktop, but your email is stored in the Cloud. In short, it's stored on a server and you see your new and saved email only when you're online.

If you have POP3 email, you can sit on the beach with no Internet access (or mobile phone service) and read all the email — and all the attachments — you already have on file. Typically, however, with IMAP email, if you have no Internet access then you would have no access to the email (and all attachments) you stored in email file folders in the Cloud.

The Cloud also matters if you're involved with a website that uses a Content Management System (CMS). Most websites live in the cloud. In other words, they are hosted on off-site servers accessed over the Internet. Some CMS systems, however, are more Cloud-based than others. Compare Adobe Contribute with WordPress.

Adobe Contribute typically works like this. First, the site is usually built using Adobe Dreamweaver. A copy of the site is then saved on a local computer — an on-site computer — and also shared over the Internet through Web hosting. People manage the site content using Contribute, which is a software program installed on their individual computers.

WordPress, on the other hand, is Cloud software. Unlike Dreamweaver or Contribute, WordPress software lives in the Cloud. Its powerful CMS lives on the site itself. With WordPress, there's no need to install software on a computer.

Edit in Contribute and you (temporarily) have a copy of the website page on your machine. Edit in Word-Press and the page you are editing is in the Cloud. In Contribute, if you save a draft of a page, you save it to your machine. In WordPress, you save a draft to the Cloud.

Now, consider the example of the organization suffering from a stolen file server and back up files. Many small businesses use a file server to manage computer files and back up all those files, perhaps every night, to another computer or external hard drive in the same office. The original files, and the backups, are not in the Cloud — they are right there in the office.

Some businesses, however, run their nightly backups using a Cloud-based backup service. New files get copied (updated) over the Internet to a server in some other location. Backing up using the Cloud, assuming the process is set up properly, definitely offers greater protection. Fire, theft, a security breach and a lighting strike are four examples of disasters that could strike a single location and wipe out both the original files and the backup files.

Cloud computing extends to dozens of other areas, but these three — email, website CMS and file backup — help show what the Cloud is and why you should care. Whether you plan to get away from it all on vacation, or thieves get away with all your computers, the Cloud can make a huge difference and help give you a sunny outcome.

Chapter 4

Comments, Reviews and Ratings

Ethics Trashed, the Law Broken in Ratings and Reviews Sites

A couple of months ago we discovered a use of online reviews that shocked us. To retain anonymity, suppose our client was an orthodontist specializing in serving children and teens. In completing routine Internet marketing work, we were surprised to discover a negative review and rating of his practice in an online reviews/ratings site. On this site, the reviewer, Mr. X, had written three anonymous reviews. We read them all to get a sense of the reviewer.

In one Mr. X raved about an orthodontist who competed with our client. In the third review, Mr. X raved about his website developer. This particular site identified reviewers by their email address, so we used the email address to search for more information about Mr. X. Imagine our surprise when we learned that Mr. X's email address matched one used by the competing orthodontist. Through this email address, that orthodontist slammed our client and gave himself a rave review. We quickly determined that the orthodontist's email address was also used to give his website developer a rave review. That suggested to us that perhaps the orthodontist himself wasn't taking these unethical and illegal steps. His website developer was.

The Federal Trade Commission (FTC) has already ruled that this kind of practice violates the law. The reason is simple. The FTC calls it "Truth in Advertising." Truth in Advertising includes, the FTC says, includes its Endorsement Guides, which state that "if there is a connection between the endorser and the marketer of a product that would affect how people evaluate the endorsement, it should be disclosed." The FTC's guidelines also state that "marketers who are compensated to promote or review a product should disclose it."

Dentists, doctors, lawyers, CPAs — professionals of all kinds — already face major challenges in identifying and monitoring online review and ratings services. For example, consider the challenge physicians face. People can read and write physician reviews and ratings in dozens of places, including Google Maps, Yahoo! Local, Wellness.com, HealthGrades, UCompareHealthCare, Insiderpages, Manta, Vitals, Everyday Health and more. In sampling these directories with just one local physician's name, the directories collectively provided five different office locations for one doctor. Four of the locations given were wrong. One was even in the wrong state! Equally bad, more than one directory was incorrect in naming the medical school from which the doctor earned his degree. Two directories listed the wrong area of practice.

Lawyers face the same challenges, with online directories including Lawyers.com, FindLaw, Yelp, DexKnows, Superpages, AttorneyLocate, Justia, Avvo and more. These kinds of "guides," complete with ratings and reviews, are relatively low in cost to program and launch nationwide, so it's likely their numbers will grow in the coming year or two.

A consumer can find it difficult to judge which site is trustworthy, which one not. A consumer may not realize that some of these directories are mash-ups — they are created by bots that scrape websites for data and then compile content from a range of sources with little or no regard for accuracy.

What's a professional to do? One professional initially told me, "I really have never thought about what's out there [on the Internet] and I don't want to." Ignorance is not bliss. In fact, ignoring these directories is a bad choice because the misinformation there can unfairly damage a person's professional reputation and confuse prospective new clients, even if they're just trying to find the office.

One step professionals should take is to Google themselves and review every single organic link in the first 10 pages of results. Many professionals who do this are in for a surprise or two. They may find old information, misinformation, incomplete information and confusion created by two people with similar names.

Here's another step to take. Google the relevant

category of service, for example, "accountants Tucson" or "personal injury lawyers Tucson." A generic search like this may identify some directories or "guides" that a search by first and last name does not turn up. Each directory or listing represents both a risk and an opportunity.

Most directories and guides are here to stay, and some, such as Google, Yelp and Yahoo!, may have a significant impact on search engine optimization and on the acquisition of new clients. It's best to attend to them all.

Most guides allow listings to be claimed and edited at no cost. It's even possible to enhance search-engine ranking by adding images and details to these listings. Another good practice is to keep a master list of directories and review it regularly. In dealing with reviews, work proactively to get positive reviews and try to get negative ones deleted.

Professionals can no longer afford to rely on word-of-mouth reputation. That reputation, no matter how strong, can sustain serious damage if online directories and guides are ignored or neglected. Professionals should complete a review-and-maintain process at least every quarter to enhance search engine ranking and protect professional reputation.

Hiring a marketing professional to complete this online due diligence often makes sense. But the marketer's work must be ethical and legal. No professional wants to have to deal with an ethics complaint, much less with a call from the FTC.

Chapter 5

Domain Names

Choose Domain Names the Indiana Jones Way

In the movie Indiana Jones and The Last Crusade, the villain chooses which cup is the Cup of the Covenant. His choice, a golden chalice covered in jewels, as it turns out, kills him. He ages incredibly fast, dies grotesquely, then turns into dust on the floor. The knight observes, "He chose poorly."

Choosing domain names poorly won't turn you to dust, but certain choices can protect your trademark and/ or business name, enhance your search-engine ranking and make it easier for people to remember your website address.

Choosing well gets more challenging every day, as thousands of domain names are purchased daily and domain name re-sellers watch like vultures for domains to expire (sometimes accidentally). Then, and then they swoop in, buy them up and offer to sell them back to you at 10 times the price.

A few of the tips conflict with one another. For example, on one hand it's great for Search Engine Optimization (SEO) to have key search words in your domain name. However, it's also true that changing the primary domain name on a website with a high search engine ranking can damage that ranking, even if all three keywords are in a new domain name.

1. Learn the Lingo

Primary Domain — You can use more than one domain name. Your primary domain is the one that your website resolves to — the domain that shows up in the address bar when the site opens. A secondary domain can and should be "forwarded" to your primary domain, unless you are planning on using secondary domains as alternative landing pages.

Top-Level Domains (TLDs) — "Top-Level-Domains" include the familiar .com, .org and .net along with .biz, .co and even domains identifying a country of origin, e.g. .uk for the "United Kingdom."

Subdomains — A subdomain is a subordinate domain that's based on a primary domain. Suppose the primary is *XYZCorporation.com*. A subdomain focusing on employment could be: *jobs.XYZCorporation. com*.

Registrar — A registrar is a company that has acquired

the right to sell domain names. There are thousands of registrars in the world.

2.   Make Dot.Com your first choice.

Unless you are a nonprofit, stick with .com as your primary domain. If you try to use .net, you run the risk of people automatically typing in ".com" and winding up at someone else's website.

If you are a nonprofit, buy the .com version of your domain name and use it to "forward" people to your .org address. See the comment above about people typing ".com" automatically.

3.   Protect Your Brand and/or Your Name: Buy .org and .net

If they are available, own both the .org and .net versions of your website. If you can afford it, buy up the .biz and .info versions as well. That way you prevent competitors from trespassing on your space, or worse.

4.   Acronyms Save Space on Business Cards, but...

If you are convinced that the target audiences you care about know you as ICHR (for short), then sure, it's great to have *ICHR.com* as a short domain name. That fits well on a business card. But if you think people will search for you using the phrase "International Consulting in Human Relations," then you just figured out what your primary domain name should be.

5.   No Worries — Domain Names Are Not Case Sensitive

Domain names are not case sensitive (neither are email addresses), so it's a good idea to use caps to make your domain easier to read and match your logo and legal name, e.g. *CenturyLink.com.*

6.   Watch Out for Unintended Meanings

You may be thinking that your organizational name, "Experts Exchange," will make a great domain name. However, someone might see that domain like this: *ExpertSexChange.com.* So just pay attention.

7.   Buy Your Name as a Domain

If you're self-employed or your name's a part of the business, then protect your personal brand by buying the domain "*firstandlastname.com.*" Type "*DaveTedlock. com*" into the address bar on your browser and you'll go to *NetOutcomes.com.* There may be a few hundred people named Dave Tedlock in the U.S.A., but only one of them owns *DaveTedlock.com.*

8. Long-term Expiration Dates Improve Search Engine Ranking

Dozens and dozens of factors determine a website's search engine ranking, and one of them is the expiration date of the domain name. If you're not a fly-by-night firm and you can afford it, renew your primary domain for at least 10 years right now. Long-term renewals also save you money; a 10-year renewal can cost you $10, or less, per year. $100 for better SEO, not to mention an accidental nonrenewal headache, is worth the money.

9. Ignore Renewal Scams

Chances are you've already received email and postal mail scams telling you that your domain name is expiring now and you must renew. One client of ours received a red-white-and-blue, official-looking document announcing an imminent expiration when in fact the renewal date was three years away. The scam includes not only a lie about the expiration date but also a huge hidden fee when you want to transfer that domain name to a reputable registrar.

10. Keep a Hard Copy Record of Your Domain Names

One way to deal with expiration notices is to keep a one-page document that lists your domain names, their expiration dates and the registrar. That way you'll have your own, independent source for fact-checking.

11. Domain Names Should Be Registered through Your Own Organization

If you let them, some website developers/technology firms will register your domain names in their name, so they own them. You wouldn't let them buy a house or a car with your money and then put the title in their name, so don't let them do that with a domain name.

Bottom-line, people want to own your domain name for their own financial reasons. One reason is that the vendor may have a wholesale account in which the hundreds of domains in it cost the vendor $5 a year while you're charged $35. Another reason is control: domain names are sometimes held hostage over a billing dispute. One hostage crisis NetOutcomes successfully resolved required us to track down a developer who had moved twice and then persuade him to accept a substantial cash payment so that the company could go back to using its own name again.

12. Use a Valid Email Address in Your Registration

Your registrar will email you when your domain name needs to renewed. If the email address you originally provided to the registrar is no longer valid, you'll never get the reminder notices. We've been called by quite a few organizations who can't understand why their website is down and their email quite working — they just didn't realize they had let their domain name expire.

13. Transferring to a Cheaper Registrar May Cost You Money

When you buy a domain name from any registrar, you are required to agree to terms that nearly always include a "transfer fee" if you decide to change registrars. The fee may be equal to two or three year's worth of annual registrations.

14. Bait and Switch Lives Here

Some registrars acquire new customers by offering you a "free year" when you buy or transfer your domain name to them. These registrars get their revenue back by charging you huge renewal fees, or huge transfer fees if you don't renew.

Chapter 6

Email Campaigns

Smart Marketing Makes Email Campaigns 'Responsive'

Websites Today Must Be 'Responsive'

Digital marketing now requires us to build "responsive" websites — websites that identify the device the visitor is using (smartphone, tablet, computer) and respond by delivering a website appropriately designed and configured for that device. Responsive sites are now vital because as much as 25 percent of all website traffic is coming from smartphones. Responsive websites also save time and help ensure that website content is up to date. That's because, by definition, a responsive site lets an organization manage all the site content in one location.

Responsive Email Campaigns Represent a Greater Challenge

Responsive email campaigns may be even more vital. Here's why. The use of smartphone email is already so widespread there's a 50-50 chance your email campaign will be opened on a smartphone before it's opened on a computer. Studies show that tens of millions of people in the United States read their email on their smartphones before they even turn on a computer. This use of smartphones challenges marketers because of the huge screen-size difference between smartphones and computers: about 320 pixels wide for a smartphone, 600 (viewing email) for a computer.

Responsive Email Produces Appropriate Formatting

A "responsive email campaign" identifies the device being used to open the email and delivers an email sized and designed for the device. Makes perfect sense, right? Here's the major problem. Even leading providers of email campaign services, such as MailChimp and Constant Contact, are just now beginning to deal with the Smartphone-Email = Responsive Email Campaign equation. It appears that MailChimp is ahead of its rivals in responding quickly and fully to this challenge.

Joel Hughes, a Senior Vice President at Constant Contact, points out that the company has added mobile-oriented services including "... Scan-to-Join and Text-to-Join options, which allow new contacts to sign up for customers' email lists via QR codes or SMS text." However, if you can't figure out for yourself how to make a Constant Contact email campaign responsive, the only help you'll get from Constant Contact is access to its "Market Place," where you can try to find a vendor to help you.

MailChimp, on the other hand, has templates we were able to use, in our testing, that did produce a fairly effective "responsive" email. The email looked great on the personal computers as well as on the iPhones and one Droid we tested it on. We found, however, that one "older-version Droid" user and two BlackBerry users reported problems. MailChimp's template may not be perfect, but it's a great start. In fact, MailChimp promotes its smartphone services right on its home page, which is more than can be said about more than a dozen of its competitors that we sampled. Among those, several email campaign providers did not even understand what we meant when we asked about "smartphone email," and none understood the term "responsive email campaign."

The reporting of the results of email campaigns remains disappointing, even from brand-name providers such as Constant Contact and MailChimp. If you send out an email campaign, wouldn't you like to know how many opens were from a smartphone and how many from a computer? That kind of report could change future email campaigns. Today, however, it seems neither Constant Contact nor MailChimp have developed campaign reporting tools that distinguish between personal computer opens and smartphone opens.

The extraordinary boom in smartphone use is opening doors of opportunity for those who want to take advantage of it. Responsive websites and email campaigns are two great opportunities. Whether we are responsive to these opportunities is up to us.

Research Proves It: Email Helps Get, Keep and Regain Customers

Q*uiz time.* Can you name the two, medium-size orga-nizations described in the case studies below? These two — called ABC and XYZ below — showed creativ-ity and courage in implementing email campaigns to get closer to their prospects and customers. That's right. Email.

Details. Organization ABC finds itself unable to use its own facilities to solicit customers during a crucial sales period. To be specific, ABC's own telemarketers won't be able to use its phone system or offices. What does ABC do? It goes with an email sales campaign. Results: ABC increases revenue by 383 percent during the one-month period compared to the previous year. How would you feel about a 383 percent increase in revenue, especially given the limitations above?

Now consider the success of the XYZ organization. XYZ decides to send a monthly e-mail newsletter to half of its prospects and current customers. The other half will not receive the online newsletter. XYZ's approach to the email campaign is relationship selling. The goal of the email newsletter, called @XYZ.com, is to strengthen relationships with those prospects and customers. A side benefit is that there are no printing, handling or postage costs.

XYZ compared revenue results from those getting @XYZ with those not receiving it. Consider these results.

First, XYZ attracted 8 percent more new customers from the prospective customer group that received @XYZ.

Then, what about current customers? XYZ found that sending @XYZ to current customers increased retention rates by 5 percent. Now consider what happened to former customers? Sending @XYZ to past customers caused 10 percent more of them to become customers again.

In short, both the ABC organization and the XYZ organization increased their revenues through the use of email. Both documented their successes and shared them publicly. Who are they? These savvy marketers were Wake Forest University (the ABC organization) and Stanford University (the XYZ organization).

Who were the prospects, current customers and past customers? Alumni. Now, your first thought may be to discount these success stories in various ways: "Stanford graduates are rich." "People automatically give to their alma mater." "They're nonprofits so they don't matter."

Whatever disconnect you are having from the reality of these successes, stop yourself for a moment and get real. The fact is that nonprofits are nearly always more challenged than for-profits in finding ways to increase revenues, particularly when technology and its expense are involved. In other words, if Wake Forest and Stanford can do it, most for-profit companies should be able to as well.

You may be thinking, "Well, sure, those two organizations were early adopters, but today so many people send out e-newsletters that no one wants to read them." That's just not true. What is true is that you have to make your newsletter worth reading. You have to make it pay off for the reader, and then get it to the right target markets. Do those two things and you can be the ABC (or the XYZ) of your marketing niche.

Chapter 7

Internet Marketing

A CPA, a Lawyer and a Surgeon Take the Helm at Online Marketing

Imagine a Certified Public Accountant (CPA), a lawyer and an orthopedic surgeon each taking the helm as the online marketing director for their multimember practices. Professionals, including accountants, architects, engineers, lawyers and physicians, typically have strong opinions about most topics, so imagine three of them making the online marketing decisions.

Your mission, should you decide to accept it, is this. Review the three opinions expressed below. Then decide which opinion belongs to which professional. The statements might be made by an architect, an engineer and a physical therapist, but here you just have three choices: CPA, lawyer or surgeon. Imagine each representing his or her practice as a whole. The word "customer" is used below for two reasons. One is to keep the focus on what's essential for success: the customers. The other reason is not to give away a correct answer — only the surgeon would call customers "patients." Ready?

Professional 1: "We don't want the kind of customer who would choose us because of our website. We just want current customers to be able to find us using the Internet."

Professional 2: "Prospective customers are just going to use our website after they've already been referred to us. We're not going to get any customers just from Internet marketing."

Professional 3: "We want our website and all online marketing to bring us more customers. We know the website's already helping us get customers, and we want more."

These three opinions about the marketing of professional services online probably represent the most commonly held views on this subject. Sure, there are other types of professionals when it comes to Internet marketing. The "Pretend-It's-Not-There" professionals do not even want to consider what the Internet communicates about them. These pros haven't even Googled their own name. On the other end of the spectrum, another group of professionals is wildly enthusiastic about social media. Let's set aside these minority positions for another discussion.

Professional 1, above, no doubt believes in the traditional approach to marketing professional services. Prospects "ask around" before making a phone call to a professional. Therefore, referrals — from someone the prospect knows — are everything. In its most fundamental form, professionals who use this marketing model have three tactics. They "put a shingle out," concentrate on doing great work and are respectful of the professionals who can and/or do refer customers to them. They believe that if they execute these tactics, their practice will flourish. Thus success depends on the quality of your work and who you know.

Professional 2 has a different perspective. Here the belief is that while all good clients do come from referrals, a website must facilitate and finalize the process whereby a prospect becomes a client. Preferably the website has two functions. It helps screen out inappropriate prospects. And a website turns a referred prospect into an appointment and first meeting.

Professional 3 has a totally different perspective. This professional believes online marketing can cause customers to choose his or her practice above others — and traditional word-of-mouth marketing is not necessary. The belief is that customers are empowered to make choices, even about professional services and some customers make a choice without asking for or getting a referral.

Put differently, the belief is that people go online, study up on services offered, evaluate credentials, and decide based on many factors, including the quality of a website. In this model referrals still help, but they live online. Research shows, by the way, that people are influenced by online ratings, even when the opinions expressed are from complete strangers.

Now can you say which professional believes the

practice gets business directly from Internet marketing? The CPA firm does tax work for both wealthy individuals and for a wide variety of organizations. The law firm serves both individuals and businesses. The orthopedic surgeons do the full range of surgery that orthopedic surgeons do.

Surprise! It's the surgeon who believes online marketing has a direct impact on how many patients choose him as their surgeon. Of course primary care physicians are influencers — they typically make a referral to a surgeon — but our surgeon explains, "Physicians don't refer patients to just one surgeon anymore. Patients want a choice. Their doctors give them choices, and patients use the Internet to help them choose their surgeon."

In this example, it's the CPA who doesn't want clients who come from the Internet. The lawyer, on the other hand, says he's sure lawyers get checked out online, but prospective clients decide after meeting the lawyer.

Did you get all three answers right? The real question is not who's who here but what do you believe Internet marketing can do for you? If your answer is "little or nothing," you will most likely make the choices required to meet your own expectations. Instead, you could choose to expect more from Internet marketing, and get it.

Blur: How Technology Confuses an Organization's Decision-Makers

In their insightful book *Blur: The Speed of Change in the Connected Society*, Davis and Meyer focus largely on the ways in which technology blurs the lines between home and work, personal and professional. Computer and Internet technology has spawned another kind of blur, one that can seriously damage organizations. Bad decisions are made in technology and marketing/communication (marcom) about who should do the work and make the decisions. The blur comes at the intersection of the technology that enables communication and the experience, talent and skill and required to communicate.

Now, put yourself to the test. Read the minicases of blur here and decide whether the responsible person is a technician, graphic designer, writer/editor or a marcom (a person overseeing all marketing and communication).

Warm up on this easy example. A communications coordinator can't send email that is vital to her work. She gets an error message in Outlook. Who's responsible for tactical success here? Right, a technician. It turns out the SMTP port on her computer must be set to 2525, not 25. The email's ready to go, but a technical error has it on hold.

Next case, perhaps more difficult. An organization has an established logo. The logo is made up of a graphic treatment or "icon" alongside a specific type treatment that spells out the organization's name. Think "Bank of America" as an example, with a U.S. flag as its icon. On the organization's website, the logo used is not an approved version. Why? "That logo didn't look right on the website. It didn't work."

Was this a technician's decision? A graphic designer's? A writing/editing decision because the "website logo" broke the organization's name in half, placing half the words on each side of the icon? Or was the logo change a marcom decision?

Technically, the approved logo would work well on the website. At the simplest level, then, the decision to use a new logo was graphic. More importantly, the senior person involved made a marcom decision. A major change to a logo is a change in branding. Brand management is the epitome of marcom.

Here's a different case. An organization successfully uses an email campaign to reach a high percentage of its main audience. One person writes the message. Another uses Constant Contact or MailChimp to send out the email. The email uses a design template that is art directed by a graphic designer, and the html is coded by a programmer.

Suddenly, the organization makes a major change in the email campaign. A new message must go out about a new service. The content comes from a different writer. The email's subject line, image and links must be changed. Who needs to be involved?

Well, the technician must send the email. A new image about the service means a graphic design decision must be made. Ultimately, a communication manager must oversee the whole process, making sure the subject line, written content, images and links make sense — and work together to promote the new service.

Next case. An organization understands that having 500 words or more of the copy on the right pages is one of many (perhaps 101) Search Engine Optimization (SEO) factors. The organization dutifully produces more than 500 words of copy for key pages. Who is involved? A technician runs keyword analyses based on search data and visitor reports. A writer provides the copy because the site is for actual human beings. A graphic designer provides the look of the page. A technician writes the code for the site pages. A marketing or communication manager should manage the process, oversee the decisions and examine the results.

It is almost impossible for one person in an organization to be a highly competent and talented technician, graphic designer, writer and marketing manager.

Some organizations blur these lines without knowing it. The result is an email, website, online ad, or other communication product that fails in one or more ways. Organizations also blur these lines because of economics. The cost of having all these talents and skills on staff is too high and the organization doesn't want to pay vendors for them.

In *Blur*, Davis and Meyer advise us to quit fighting the blur between personal and professional and capitalize on it instead. The same advice applies to the blur involved in the roles of technicians, graphic designers, writers, editors, creative directors and marcom managers in our technical world. Each organization must identify and be honest about the talents and skills of the people involved, take advantage of the synergy and rely on a communication professional to keep the organization on goal, on strategy, on tactic.

Imagine a new campaign with a new, simple message. A bumper sticker, a poster, a Tweet, an email campaign, a Facebook comment. The teaser: Blur Happens. The campaign theme: Blur Happens. Make It Work.

17 Ways to Sell Green Eggs and Ham

The economic environment today is so tough and so competitive that many businesses and nonprofits wonder whether enough customers are ever again really going to buy, or buy into, what the organization is selling. Some might say business is so bad that instead of selling, say, a stunning floral arrangement, customers act as if the organization is selling something ugly or even repulsive, something like green eggs and ham.

Yes, I'm talking about *Green Eggs and Ham*, by Dr. Seuss, a book that is essential reading for anyone concerned about sales and marketing. *Green Eggs and Ham* offers a couple of valuable lessons, particularly when it comes to Internet marketing. These teachings can be applied to any organization and its marketing approach.

The first lesson from *Green Eggs and Ham* comes from the first comment that the sales person (Sam-I-am gets from his prospect. When Sam first asks for the sale, the prospect says, "I do not like that Sam-I-am." Worse yet, the prospect looks angry. And his fist is clenched! Sam-I-am keeps right on selling though, and remember, Sam is selling, of all things, green eggs and ham. The lesson: don't give up on a sale even if the prospect doesn't like you or your product.

Be like Sam-I-am. What does he do when he is rejected? He tries more than 20 times to make the sale, in 17 different ways. Really. Get the book and count for yourself. First, Sam-I-am tries to sell green eggs and ham by just asking for the order. Then he tries three variations: here, there and anywhere.

When these don't work, Sam-I-am tries a wide range of selling scenarios, ones involving a house, mouse, box, fox, car, tree, train, dark, rain, goat and boat. Eleven more times he asks for the order.

In the middle of this creative selling spree, Sam-I-am stops and simply asks, twice more, for the order. Finally, his 17th sales effort works. For his 17th attempt, Sam-I-am goes with the time-honored trial-use offer: "Try them. Try them." He gets the sale. Better yet, the once-angry prospect, now a customer, actually thanks him for being so persistent.

Unfortunately, only a handful of highly proactive leaders in organizations take the Sam-I-am approach to marketing and especially to Internet marketing. These innovators use every tool they can to create relationships, create prospects, retain customers and close on sales. Their effort is relentless and protracted. If there's a situation or way to sell, they find it.

Consider taking the Sam-I-am approach to Internet Marketing for your organization. In honor of Dr. Seuss, here is Sam-I-am re-cast as a digital marketing and sales person:

Do you like our website pages?
Have you put us off for ages?

Does our website meet your needs?
Do you want an R-S-S feed?

Does our rank on Google help us?
Should we make an S-E-O fuss?

Do you check out Google places?
Have you seen our happy faces?

Do you like our Google ad words?
Are we using the right adverbs?

What about Yahoo! and Bing?
We can give your sales some zing.

Do you like our Yelp reviews?
Have you seen us in the news?

Would you like a promo email?
Do you want our bloggers female?

What about a YouTube show?
We could start a blog, you know.

Would you like us on your smartphone?
Is our pitch right in your buy zone?

If we Tweet you and you Twitter,
We can give your life some glitter.

Would you like us right on Facebook?
Have you Stumbled Upon our look?

Would you see us right on Craigslist?
Tell us what it is we've missed!

Try us, try us, you will see
Online is the place to be.

As you can see, the digital marketing Sam-I-am may need to provide 18 ways to create a relationship and ask for the business. And ask and ask.

The version of *Green Eggs and Ham* I keep in my office says on the top corner, "I can read it all by myself Beginner Books." If you are a beginner, you can still try digital marketing all by yourself.

In my opener I compared selling green eggs to selling a stunning floral arrangement because a florist I know practices digital marketing every day. He is the Sam-I-am of selling flowers through Internet marketing, and his efforts keep him in business while other florists falter.

True, life in the real world means, unlike Sam-I-am, you probably can't try every digital marketing option out there. What's more, everything you try doesn't have to be digital — maybe you need to mix in an event, try some print advertising or become a sponsor. You can get professional help making choices, but Dr. Seuss and I are giving you three clear-cut pieces of advice about your organization and this economy. One, don't give up, even on unfriendly prospects. Two, create new places and new ways to build relationships with prospects and current customers. Three, ask for the sale, over and over and over again. Seventeen times should do.

Chapter 8

Managing Geeks and IT Staff

What Type of IT Person Do You Have?

In meeting hundreds of CEOs and IT managers, I've seen top management struggle more with hiring and managing IT people than in any other area. Increasingly it seemed to me that IT people came in a variety of types. More importantly, by knowing the types, CEOs and managers from other departments could work with them more effectively.

To test my theories, I had discussions with three experts in IT: a highly experienced corporate IT consultant, a small business IT consultant, and an in-house VP at a medium-size organization who managed an IT department among many others. As our talks progressed, the kinds of IT people we identified grew. As you read through these descriptions, think about the IT people you know and ask yourself where and whether they fit into one of the categories below.

The Pretender. Secretly incompetent. Hides behind geek

speak. Can't explain IT in a way people understand. Projects take much longer to complete than anticipated. Ultimately gets found out, but still represents a genuine threat to the organization. One Pretender I warned a CEO about allowed the website to be hacked, allowed vital files to be deleted when a second hack occurred and had failed to set up a backup system, so thousands of computer files were permanently lost. If you suspect you've got a pretender (on the payroll or under contract) get a second and even a third opinion and find out before it's too late.

Mr./Ms. Indispensable. Keeps no documentation. Makes every project more complicated and unique than necessary. Secretive. Not willing to explain existing systems, share passwords and more. Creates serious business continuity and security issues. Potentially as dangerous as the Pretender — if he/she leaves, no one has a clue about what's going on.

The Theorist. Thinks and works from the 50,000 foot level. Has great, big ideas. Plans well. May outsource too much. Can't get tasks and projects completed. In a large organization, a theorist may be useful. In a very small organization, a theorist is often useless.

Mr./Ms. Self-Taught. Learns by doing. Task-oriented. Completes some projects. Struggles each time a need in a new area arises. Doesn't know what he/she doesn't know.

Rarely plans, may struggle to manage. Organizations may try to save money with the Self-Taught, but limitations and liabilities come with this choice.

Mr./Ms. Details. Gets some projects completed, but doesn't plan or manage resources (human or money). Is surprised by new demands. Highly task oriented. Fails to prioritize. May work long hours and burn out.

The Resume Builder. Always planning for the next job, not according to the organization's needs as identified in its business plan. Spends the organization's money on training, seminars and special projects that build the resume. Tends to use acronyms to sound extra competent.

Not My Real Job Here. Got assigned IT because no one else would do it. Real job is in HR, Marketing or elsewhere. Management has to recognize the limitations of this approach and the low commitment that results.

The Super Hero. Always answers, "I can do that." Is highly productive. The "utility infielder" of IT. Works long hours (which nearly always pleases management). Takes on all manner of projects. Completes most. May be vital to a small organization. Even so, misses some deadlines because the Super Hero actually could do it all, if only there were 30 hours in a day. Often burns out.

The Go-to-Guy. Great to have at any level. Good communicator across departments. Can explain IT issues in plain English, so people go to him/her for solutions. Knows how to get things done. Is sought out by other managers because of the ability to get projects completed. Excellent problem solver. Many organizations are lucky to have just one Go-to-Guy in IT, or in each department of IT.

The Eclectic. A Go-to-Guy to the second power. "The Eclectic" has a wide range of skills and experience, often in unrelated areas. Can explain complex IT issues in plain English to anyone. Can plan and manage both budgets and people. A great problem solver, and not just in IT. Has worked outside IT, typically in more than one area — HR, Marketing, Accounting, etc. Proactive. Understands that the organization's business plan must drive the IT plan, and makes decisions accordingly.

If you've got a Super Hero, Go-to-Guy or Eclectic in your organization, count yourself as lucky and don't brag about it — he/she might get recruited away by someone else! If you have one of the other types, take steps to do better. You can.

Tips for Dealing with IT People

Most marketing people are, at times, frustrated by the Information Technology people they have to work with. At worst, marketing people describe their IT counterparts as difficult, obstructive, uncommunicative and even incompetent. Yikes!

Given the sometimes great divide between marcom and IT, here are some tips for business owners, entrepreneurs and marketing people on getting what you want from your IT people.

First, remember that what you want is probably just one item on a long list of demands being placed on an organization's IT person or staff. Here's an example. One trip took me out-of-town, to corporate headquarters, where I was scheduled to meet with the head of one of the divisions of a company and then its CEO. I knew the company's solo IT person, Steve, was highly capable and personable. He met me in the lobby and my company tour started.

First, we entered the offices of a division of the company, a division with its own, terrible website, Steve told me. The moment Steve stepped into the lobby, he was accosted by people wanting help. One woman had a problem with her desktop computer. Another was having trouble with the phone system. A third wanted help figuring out how to use her new company cellphone. When we finally made it into the VP's office, the VP ignored me and spent the first 15 minutes of our visit making Steve show him how to perform basic functions on his cellphone. When we finally got around to the marcom / technology issues, the division head had just a few minutes left for us — but he was now able to use his smartphone!

Second, realize that the technology problems an IT department or person must solve are surprisingly unpredictable. Sometimes a complex problem can be solved quickly and easily. At other times, a relatively minor problem can take hours and hours to solve and put some other project behind schedule. Sure, the best IT people plan accordingly, delegate and create appropriate expectations. Even so, IT problems can confound experts, inside the company or outside of it. Honestly, sometimes even a solution that works can't quite totally be explained. It just works.

Next, take some responsibility for educating yourself. Whether your issue is buying a new domain name, improving SEO, getting a better cellphone or making voice mail on your phone system work, do your homework first. Find out what you can about the situation you are in, how it's supposed to work and what you think the solution might be.

Do not start a conversation with an IT person by saying, "I just don't understand technology." Seriously. If you've already made up your mind that you don't understand technology, how is an IT person supposed to explain anything to you?

You can also help yourself by providing as much detail as possible about the problem. If you call the Help Desk and simply say, "My computer's doing this weird thing," that's a bad way to start to the conversation. If it's a website page that's a problem, learn how to take a screen shot of that page (hold down the CTRL key and press the Print Screen key). If there's an error message on that page, a screen-shot will capture that information. Suppose, instead, you think you're missing some emails. If so, be ready to provide the email addresses of the person or people who say they sent you email that you did not receive — or lost.

Lastly, learn some lingo. Technology people are guilty of geek-speak, no question. It's equally true, however, that technology requires new terminology to be invented every day. Sure, it may be annoying to have to remember that Barracuda is a service that's providing a spam filter for the company, or that the company's using a Windows server for its website, but if you get your language right from the beginning, your IT person will be less distracted by your terminology and better focused on solving your problem.

All that said, you also have to watch out for IT people who use language as power. Some IT people like the idea that they can speak a language you may not understand at all. Don't let them. Tell them to speak in plain English and keep telling them to do that until they do. Any competent IT person who really understands a problem can explain it in plain English. If you don't get plain English, that may be a sign that your IT person doesn't understand the problem, much less have a clue what the solution is.

Lastly, get help when you need it. Sometimes IT people get stuck on a problem and just can't come up with a solution. Individual human beings are not necessarily able to solve all problems by themselves. If the solution's important, an outside consultant could be the answer.

Bottom-line: take up as much slack as you can in the problem/solution process and you'll get better results. That'll make you — and your IT person — happier.

The IT-PR Team: Who's on First and What's on Facebook?

Consider a Board of Directors meeting I attended back in 2001. A Board member had a concern about email campaigns. "I'm not sure," he said vaguely. "Our Computer Network Administrator should be managing our email campaigns."

The Chairman of the Board suggested email campaigns were technical, so having a technical person manage them made sense. In that same organization, an engineer was responsible for website strategy, design, navigation and content. Years before, when no one else had a clue, the engineer built the site and had managed it ever since.

This organization communicated strictly with consumers, with ordinary people, but the Internet's peculiar new intersection of technology and communication confuses leadership in many organizations. Some leaders have made uninformed and costly decisions when they assigned responsibility for the ever-changing field we call "digital communication" or "Internet marketing."

Emerging technologies continue to blur the lines between Information Technology (IT) and public relations (PR). The ever-present danger is that leadership will confuse the role of managing the technology used to deliver the message with the role of deciding what the message is and to whom it should be delivered. What guidelines can leadership use to assign responsibilities?

Consider these simple definitions of IT and PR, beginning with a paraphrased definition of IT from About. com. IT, also known as Management Information Services (MIS) or Information Services (IS), focuses on the use of hardware and software to manage information. IT is responsible for storing, protecting, processing, generating, transmitting and retrieving information.

The Public Relations Society of America (PRSA), never at a loss for words, spends an entire page defining the role of the PR professional, but here's a statement from PRSA's website: "The [PR] practitioner uses a variety of professional communication skills and plays an integrative role both within the organization and between the organization and the external environment."

Simply stated, IT people manage technology that processes information. PR — or more broadly "marketing and communication people" (marcom) — determines in a wide range of contexts what information needs to be shared, how it needs to be communicated and who needs to get the message. If you are saying to yourself, "Well, that's obvious," then take the test below with confidence.

For each situation described below, identify the person who should have primary responsibility — IT or marcom. Keep your own score.

1. The design of a smartphone shortcut.
2. The analysis of website visitor statistics.
3. The use and placement of QR codes.
4. The use of a text messaging campaign.
5. Content placed on a new set of website pages promoting a new member log-in service.
6. The execution of a website survey.
7. The use of domain names to improve Search Engine Optimization.
8. The analysis of an email campaign's opens, click-throughs and unsubscribes.
9. The identification of target market(s) and the design and content of email campaigns.
10. The choice of the brand of smart phones for an organization's management to use.
11. The decision to allow employees to access Facebook at work.

The correct answer to the Questions 1-9 is "marcom." Each of these situations involves technology — smartphones, QR Codes, email services, websites, text messages — but the decision centers on communication, internal or external. The answer to Question 10, about choosing a brand of smartphone, is "IT."

Question 11, about whether employees can access

Facebook at work, sometimes creates a dialogue that's right out of Abbot and Costello's classic baseball routine "Who's on first?" IT measures employee behavior, determines that Facebook and Twitter access at work seriously impacts network and Internet functionality and therefore blocks access to both. When the CEO asks marcom who's on Facebook? and what's on Twitter?, marcom may have to answer, "I don't know." Marcom can access either at work only from a smartphone. Social media work has to be done at home or at Starbucks.

How did you do on the quiz? Congratulations if you scored 100 percent! Not everyone does. In my first two paragraphs, I mentioned an organization in which an IT person managed its email campaigns and an engineer managed its website. These assignments were made even though the communication targeted ordinary consumers and marcom people were available, capable and willing to do the work.

Because I wrote "back in 2001," chances are you nodded and thought, "Well, sure, back then the Internet had most managers and leaders confused." Here's the surprise. The meeting I described actually took place in 2011.

To put a winning team together, leaders must grasp the difference between having IT manage technology and having marcom manage communication. The winning teams get the assignments right. Everyone knows "Who's on first?" and "What's on second?"

Anton Chekhov and Act 1 of Website Development: The Conference Room Meeting

In the work of the Russian playwright Anton Chekhov, his characters often ignore what each other has to say. Instead, each character has a conversation with himself or herself at the same time that other solo conversations are taking place in the room. Read an act of *Three Sisters* and you'll see.

At times when the characters in an organization meet to discuss goals for a website, it seems Chekhov could be writing their lines. Act 1, Scene 1: the conference room. Attending: the Marketing Director, the Sales Manager, the IT Director, an IT assistant and the website consultant.

Marketing Director: Well, it's more than we have budgeted.

Sales Manager: What if we developed multiple databases of prospects instead of one. That way, we'd have a wider range of options for targeting prospects by demo, right?

Consultant: You can use multiple databases with multiple autoresponders if you want to, but that approach does require a larger investment.

IT Director: I don't understand what the big deal is about sending out an autoresponder email. That's just a single line of code.

Sales Manager: But we just said it's several different responses we'd be sending out all the time. I don't even know how many yet.

IT Assistant: What did you say HR would get in terms of an online job application?

Consultant: Someone said HR wants to get a job application that looks like the hard copy application you use, plus allows an applicant to attach a resume.

IT Assistant: Why not give 'em just a box where they can cut and paste in their resume?

IT Director: We should create an employment application form that creates a database so the database can be searched for certain types of applicants.

Sales Manager: People are just going to want to send in their own formatted version of their resume, not stick it in some box.

Marketing Director: We need to ask Anna what she wants to do.

Consultant: Is Anna the head of HR? Can we get her to join us?

Marketing Director: Well the CEO called this meeting, but she couldn't make it. I don't know if Anna was invited.

IT Director: The application form should create a database. A database is the way to go.

IT Assistant: ASPX. ASPX is the way to go.

Marketing Director: What's ASPX?

Sales Manager (getting up and reaching for his cell phone as he heads for the door): I've got a call I've got to make, but I still like the idea of multiple databases.

IT Assistant: Active Server Pages.

Marketing Director: (frowning) What's that?

Consultant: One of the major languages used to program websites. PHP and ASPX are the Coke and Pepsi of programming. We can talk about the advantages and disadvantages of ASP.

Marketing Director: I don't think it's in the budget.

IT Assistant: I think ASP is the way to go.

IT Director: I still don't get what's the big deal about the auto responder.

Meetings that involve the major stakeholders in an organization's website are essential for getting the highest payoff for the site that's built. But the discussion doesn't have to seem as if Chekhov's writing the dialogue.

At least three vital elements are missing from the nightmarish meeting reported here. One is leadership — the CEO must manage a meeting like this or empower someone else to do so. A second is making sure all stakeholders are in attendance. If the HR Director, for example, is not involved in the process then it's likely that the entire HR Department won't like the outcome. Lastly, people representing various parts of an organization have to make a genuine effort to listen to and understand each other. Leadership, involvement and understanding — those are the central elements in creating a successful site.

In Chekhov's *Three Sisters*, Vershinin says, "If only we could educate the industrious people and make the educated people industrious." The best website committees, and their facilitators, should figure out how to do both. The result: an organization with a powerful, dynamic website that provides high value.

Chapter 8

Meditations on Technology

Living Our Lives Online, Not Written in Stone

The job candidate sat across the table from me and answered my question in an unhappy tone. "No, that website I worked on is no longer available. The company already changed it again."

What about her work from two years ago? Already gone. Three years ago? Gone.

True, she should have taken screen shots of these sites, but screen shots are only skin-deep site views that tell nothing about the programming underneath. At Net-Outcomes, we know all too well this problem of transient work. For example, we once created a complex website for Health Net of Arizona. Members could check the formulary, find pharmacies and choose primary care physicians, all on an appealing site that ran fast, pleased staff and members and sold services. Less than a year later, corporate wiped it out.

Consider this contrast. Our vacation home outside Santa Fe has a workshop, where, several summers ago, we finished pouring another 300 square feet of concrete floor. As my son Michael, then 12, and a neighbor, Archie West, finished troweling the final corner of the floor, Michael said, "Dad, I'm signing it now." He started this "signing" of concrete a few years earlier when we poured a fresh flagstone and concrete floor for the portal up at the house, in the courtyard. Long before then Michael had noticed the writing on the floor of the front porch, written into the concrete by my father. It reads:

Ted
Agnes
Dennis
Susan David
Happy, Missy
June 30, 1960

These are the names of my father, mother, brother, sister and me, followed by our dog and cat (at that time).

In a different section of the workshop, Michael had also noted the slab of concrete into which I'd written:

Dave
Sandy
Nicky
Cassidy
July 5, 1989

In doing so, I'd kept up my parent's tradition, naming my family at the time — me, my wife, a huge golden retriever and a very Siamese cat.

Michael, as is true of most millennials has never owned any music albums in the big, physical 33 rpm record sense. In fact, he owns far more music on his iTunes account than he has on CD. Michael uses an iPod, MacBook and iPhone to enjoy his music. But here's the key point to pondering what technology is doing to our society — his ownership is virtual, not physical. Years ago, underneath the portal, when Michael said, "Dad, we need to sign it now," he wrote our family names there, including Jazz (the dog) and Mocha (the cat), and then he hesitated. As Archie and I stood there watching him hesitate, I knew Michael was wondering about Archie.

Archie is a rancher, guitar player and highly skilled mason, a man I've known for more than 40 years, gray now, weathered with age, but still tall and strong and quick to show a generous smile below an equally generous mustache.

I said, "You can add Archie's name there. You should."

Archie immediately said, "Well now, Dave, you don't have to…" but Michael was already writing "Archie" into the fresh concrete. I couldn't imagine a better legacy for Michael than to record Archie's name there. The man is a millionaire because of the land he owns and will pass on to his son and daughter, yet he's a man without a computer, cell phone or digital camera.

For millions of people, digital cameras mean photo ownership has also gone virtual. True, some people still order prints of selected photos, but for many people their photo albums exist exclusively on a Website, computer, external hard drive or backup CDs. Thousands of photos, all virtual.

Michael's Mom appeared in the workshop on the day we finished the concrete slab just last summer, taking pictures of the finished project with her camera. Even as we stood there, watching Michael write into the concrete, no doubt thousands or tens of thousands of Websites were being wiped off the Internet, replaced by new sites or simply dumped, vanishing instantly from domain names and nearly forever from caches as Google and Yahoo! gradually forgot them and only the Way Back Machine stood between that work and oblivion.

Inside the workshop, Michael was writing into the concrete:

Dave
Sandy
Michael
Archie
Jazz
Mocha

— and a date.

I have been wondering what this means to our future — that people of all ages are leading increasing virtual lives, playing virtual games, entering fantasy football leagues, owning songs and photos only in a virtual world, knowing that their virtual work, and proof of their work experience, is temporary, certain to be erased.

I know this much. Few people write their names into concrete and then, 47 years later, have a grandson, a child they did not live long enough to even know was coming, follow that tradition and create his own physical record and sense of permanency.

I wonder what an increasingly virtual future means for us. What it will do for us, and to us. That day several summers ago, before I could reach a conclusion, I was distracted by my son saying, "Do you think we can get Jazzy to sign it? We should stick her paw in here."

I looked at Archie, and we both stood silently, pondering the difficulty of wrestling with a rambunctious golden retriever while trying to stick her paw into concrete, nearly dry now anyway.

Just then, Jazzy burst into the workshop and circled around us, leaving a perfect arc of footprints, shallow but distinct, like etching on glass, on a wide expanse of fresh concrete. Speechless, I looked at Archie and his face burst into its hundred smiles of age and miles and he said to us, "I think that's just perfect. She signed it after all."

Then we all smiled, and Sandy took another picture — with our digital camera.

A No-Tech, High-Touch Thanksgiving

You've just discovered the one piece in this book that is only about technology. And it's about going no-tech — for Thanksgiving.

Several years ago I read an article by a man who (as I recall) had survived a usually fatal disease and found himself incredibly thankful. Thanksgiving Day was coming up, and he realized he felt thankful to dozens of people who had helped him in life, and most important — he felt he hadn't thanked any of them.

His solution was to sit down and start hand-writing Thanksgiving cards that included hand-written thanks, with details. He wrote out of his own need to give thanks, without any expectations. His results were enormously unexpected. He wrote, for example, a detailed thank you to a wonderful grade school teacher of his. Just finding her address was a challenge. Finally he located her at a nursing home in another state and off his card of thanks went.

A couple of weeks later, he got a card back from her that nearly made him cry. She was alone, she wrote, having lost all of her family, and her health was failing. She was feeling forgotten — until his thank-you card arrived. It made her believe that after her 40-year career as a teacher, surely there must now be many other students who were also thankful, as perhaps he had told her in his card. His card made her day, week, month and life bright. She kept the card in her room permanently displayed, smiling when she saw it, one card having an impact 100-fold.

The year I read this story, for Thanksgiving, I started writing thank-you cards, about 75 of them that first year, thanking all the people I could think of, who are still living and whom I should have thanked already. I tried to detail how they'd helped me. Like the man whose article I'd read, I expected no response.

Now, all these years later, I have many surprises to report. One of my first thank-you's went to a man who'd shown great leadership in my son's school. More than a year later, he happened to see me in a restaurant. A big bear of a man. Construction guy. He immediately came over and gave me a hug that took my breath away and threatened to break ribs.

Then he held me back (possibly with my feet still off the ground) and said, "That note you wrote me was the nicest thing anyone ever said to me in my entire life. I feel the same way about you." He had tears in his eyes.

Another surprise for me is that each year my list

gets longer. You might think it would be enough for me to thank my friend Dave just once for being a great friend for more than 20 years. It isn't. Some people I owe a lifetime of thank-you's. Dave's one of them.

After a few years of writing Thanksgiving cards, I ordered extra cards and brought my young staff into my office. I told them the original school teacher story, and my bear-hug-and-tears story, handed them five cards each, and commanded them to write thank-you cards to five people who deserved thanks for touching their lives. "But not me," I added, in case they thought this assignment was testing them on an even weirder level than they already perceived.

They smiled pleasantly, nodded, left my office, placed their cards on their desk tops — and left them there for more than a week. Another meeting in my office. "You're not writing yet," I pointed out. "What's the problem? No one to thank?"

They looked down at their feet, up into the corners of my office. Shook their heads. No, that wasn't it. They had people to thank, all right.

Do it now, I commanded. One card. Surely you can think of one thing your mom or dad did for you. Or a teacher, a friend, someone. "Be specific. Write, 'Thanks for bringing that can of gas to me when I ran out on the Interstate at 2 a.m.' or whatever."

Off they went. They began to write. I avoided pacing up and down beside their desks.

Eventually, the five cards disappeared. I am still surprised by how difficult this is for young people, but here's another surprise. Just the other day, a Thanksgiving card arrived in the mail for everyone in our office. From a former employee, in the Air Force.

"I didn't want to stop the Thanksgiving card tradition," he wrote, "so I have kept it going by writing this one to all of you." Then he wrote two paragraphs of thanks to us about how we'd helped him in the year he'd been with us.

A couple of days ago I had lunch with a semiretired man who thanked me for my Thanksgiving card and added that just a couple of years earlier, he'd written a detailed thank-you card, at Thanksgiving, to his brother. They're both in their 70s. He wrote, he said, about how thankful he was to have him as a brother and "said things I'd never said to him before."

Then his brother wrote a Thanksgiving card back, with his reasons he was glad to have my friend as his brother. The cards, my friend told me, "changed the nature of our relationship forever."

Yesterday I received a thank you back from my former pastor, Frank, who moved to California years ago. I mention Frank here because I should point out there are some people we can't thank with a card now. Either we can't find them, or they've died. They are gone now. I think Frank would say that we can tell God why we're thankful to them, and God will pass the message along.

So my high-tech advice for Thanksgiving is to go no-tech, and to hand-write five thank you's to people you know. Hand address the envelope, too. People love that.

One woman wrote me back and said, "You're the only person who's ever sent me a Thanksgiving card with a thank-you note. I'm really touched." Now it's your turn to touch someone. Do it.

A post script. The day this column went out by email, I immediately received an email reply from my brother-in-law, who wrote, "I remember bringing you that can of gas when you ran out on the Interstate." He was happy to do so, and happy to be thanked.

Taking a Vacation from (or with) Technology

Back in the Dark Ages, when cellphones offered all the carrying convenience of a five-pound bag of sugar, my wife and I began retreating on our vacations to a profoundly energizing place. Our special haven — no doubt you have your own — in rural Santa Fe, N.M., was so remote back then we lacked not only phone service, cellular or landline, but also electrical service unless we turned on the generator.

To get in touch with our offices, we had to drive four miles to the nearest pay phone and, if it happened to work that day, slip in stacks of change and take notes as we stood in the dirt on a street that crossed the main railroad line just 50 feet away. These vacations were simultaneously awesome and awkward. We called in just twice during the week, but a genuine crisis might require hours talking on a pay phone or retrieving and sending faxes from a copy shop in Santa Fe. Otherwise, our complete escape was deeply rejuvenating.

Now, advances in technology mean that truly escaping from work requires planning and willpower. Consider a man I know who insists on taking his cellphone on vacation and then experiences great discomfort when, for example, he takes a business call while floating under Paul Bunyan on Fairy Tale Brook at Legoland.

Smartphones offer text messaging, email, full Internet access and more. Wireless laptops tie the knot even tighter. Hotels offer Internet access, and our smartphones let us stay connected on our laptops in the car for the full six-hours it takes to drive from Tucson to San Diego.

The latest copy machines can scan documents, convert them to PDF and email them to you, so you can analyze 50 pages of bank statements as your car zooms over Interstate 8's version of Death Valley at 80 mph, with your radar detector on, of course.

The portable printer remains the only technology item still about as convenient as a 5-pound bag of sugar, but you can always get the front desk to print your documents from the flash drive you stuck into your laptop.

Today, the challenge is no longer staying in touch with the office; it's how to effectively disconnect from it. Needing some recommendations myself (I still associate the music at Legoland's Fairy Tale Brook with the particularly challenging launch of a client's online job application form), I began asking Tucson businesspeople how they deal with technology on their vacations.

One man I talked to, a man who sold the very copiers that threaten to email us an avalanche of office paperwork, casually told me he never takes a laptop on a vacation and doesn't worry about having his cellphone with him because he makes sure that few people know his number.

A bank president I talked to offered five simple steps that might work for anyone:

1. Never get a smartphone. (Later, he got one.)
2. Use email from a laptop only on extended trips.
3. Retrieve messages from your direct phone line, but make sure your voice mail explains that you are out of the office and provides the names of people who are in and can help.
4. Check your cellphone for messages, but don't take it with you everywhere.
5. Breathe deeply and enjoy watching those people who let devices run their lives!

The chairman of the board of a software technology company told me he takes his vacations on Cape Cod, two weeks a year. In 18 years, he says, he's called the office just three times and they've never called him.

"Before I go on vacation," he says, "I hear, you know, 'Well, what happens if we get a $10 million contract?' and I say, 'Tell me the good news when I get back.' "

"Or they say, 'What happens if the building burns down?' and I say, 'When I get back, you can tell me how you worked it out.' "

He adds, "There's a magic to two weeks. I do my deep

thinking then, out on the bay. Vacation is a time to recreate. That's why it's called recreation. Take two weeks."

A man I know has a lovely place in rural Santa Fe.
His ticket to just one week in heaven has been a compro-
mise: install a landline, use the cellphone from the hill by
the house, and put an actual desk in the guest bedroom.

Besides, there are side benefits to taking your laptop
on vacation. For example, your child can use your laptop
from the back seat of the car to watch movies or make
a PowerPoint presentation titled, "Our Eventful Trip to
Santa Fe." Batteries, and smiles, included.

The King of Low-Tech Still Has Worries

A life-long friend of mine is lower-tech than anyone you know. He offers some lessons about dealing with technology, but perhaps not the insight you might expect. The man I know, Marc Simmons, has lived nearly no-tech for 40-plus years. Simmons lives in rural Santa Fe County, N.M. He has no iPod, no cellphone, no cable, no color TV, no stereo, no voice mail, no CDs, no computers, no Internet, no credit cards, no AC, no evaporative cooling and no central heating. There's more. He has no sink, toilet or shower — he has no running water! And no electrical service. Yet he's famous and revered in his profession.

What technology does he have? Propane for a heating stove, a land-line telephone which only a couple of years ago was rewired so he could answer the phone inside his main house, two very old vehicles and a solar-powered lantern.

He's not poor. In fact, the land he owns makes him

a millionaire. Physically, he's frail, the result of decades of time piled on top of a devastating car accident after which he died, literally, twice. Back then, his surgeons, after multiple surgical attempts to piece him back together, were still puzzled that during rehab, he continued to say that trying to walk again was excruciatingly painful. Months in, the docs re-evaluated him, ordered more tests and discovered they'd overlooked a broken hip.

In New Mexico, and among Southwestern history buffs, Simmons is famous not because of his lifestyle or his two-time revival from death, but because he's New Mexico's best-known, best-loved historian. His profession: author.

Simmons to this day writes a history column for a local newspaper. I write a technology column, so we share being columnists. Thinking of showing him my laptop one day, I cautiously asked if the many writers he knows have told him he should use a computer.

"Oh yes," he said smiling. "I hear that all the time." He writes his newspaper columns, and his books, on an Underwood typewriter the size and weight of a Mack Truck. He forthrightly tells people he uses a typewriter.

"Some of them get persistent about computer use, and then I ask them how many books they've written."

Simmons has published more than 50 books. He added, "When I tell them how many I've published, they usually get pretty quiet after that."

His avoidance of technology runs deep. He watches TV, but on a small black and white unit powered by the car battery from his Toyota 4Runner, which he parks by his bedroom. Even his hand tools are low tech. No spade he shoves into the earth has a fiberglass handle. "Fiberglass," he scoffs.

Despite escaping all this technology we take for granted, Simmons must still cope with his own particular anxieties and obstacles. He cannot email his columns to the newspaper, so he must drive to the post office and mail them. In the summer, he frets over whether a sudden monsoon rain will leave him stranded out on the highway, unable to get home over a mile of dirt road because of deep water running fast in two major arroyos that cut across his land.

The mother of all his worries, though, is his writing machine. His typewriter takes ribbons he may not be able to buy some day. To protect against a possible permanent lack of ribbons to buy, he keeps boxes of them stored in his propane-powered refrigerator. Typewriter ribbons last longer, he explains, if you keep them refrigerated.

Then, too, what would he do if his typewriter ever broke — where would he find a typewriter repair service? He has a backup machine but I suspect its keys stick, among other ailments he dreads having to cope with.

We all struggle with the problems and challenges that technology creates. We may worry about whether and when our gadgets will break or if we'll even ever learn to use them effectively.

Santa Fe neighbors for nearly a life time, Simmons and I disagree on technology matters. I think the dirt road we share requires the installation of a few dozen high-tech flood control devices commonly referred to as "culverts." But he opposes their installation as useless. On the subject of culverts, we've reached an impasse.

Until recently I was sure I was right about them. Then I became not-so-sure. In an ordinary engineering feat accomplished years ago, Pima County installed a pair of five-foot-in-diameter culverts to make an arroyo flow underneath the street that provides our northern route to work in Tucson. The culverts kept the bottom of the arroyo a good eight feet below street level.

However, in our biggest storm this year, both of these gigantic culverts filled with sand and vanished, permanently. At the same time, our arroyo abruptly adjusted its own depth eight feet upwards. Now the water in it from monsoon rains flows a foot or two high directly across the street. Twice in one year alone, major storms prevented us from leaving home to go to work.

It's true that if I had tried to cross the road, my nearly-new Toyota Highlander would have been swept away. Instead, each time, I just stayed home and worked from there. On my laptop.

Chapter 9

Mobile Marketing

A Real Estate War over 36 Pixels Squared: Get in the Trenches Now!

A war is raging right now over some of the world's most precious real estate. Millions in armies toil in the trenches. Billions of dollars, millions of jobs and the success of tens of thousands of organizations are at stake. The contested real estate is tiny in size, as little as 2.5 x 3.0 inches. A limited number of parcels are available and worth having. Each parcel has a limited number of lots, each lot the size of your finger tip, and the rent is free.

Win today's battle for space, get a lot on a primo parcel and you might get a big payoff. But your lease can be cancelled any second of any day, without notice, and your competitors range from Facebook to SeaWorld, from a local school to Arizona Athletics.

The "parcels" are the home screen, and on the additional "pages" after it, on millions of smartphones. Available lots range in size from 36 x 36 to 114 x 114 pixels, each occupied by an icon, a shortcut. Some of the most desirable locations are pre-leased. Some leases (e.g. Camera, Settings) can't be cancelled, even by the smartphone owner. Each owner decides what icons live on which page, and how many pages there are. Perhaps no marketing space in history will be more hotly contested than these precious, tiny, pixel spaces.

The original iPhone fits 16 icons on its home screen, Droids often more. Smartphones stack multiple pages after the home screen. You can be sure that extensive, private research has already been completed to project how many screens the average person will use and what an organization has to do to keep its lease there. The catch is that the companies that completed this research will never share it with you.

My guess is that the average person may juggle from three to seven home screen pages, or about 48 to 60 icons. Folders can be used to hold more icons on fewer pages, but no organization wants its icon tucked into a folder — that just makes it harder for a person to use. Every organization wants to be on the home screen's first page.

One summer my family visited SeaWorld, where I discovered, as we parked, a SeaWorld app I had added to my smartphone's home screen. The app remembered exactly where we'd parked, told us the walking distance to any ride or show and estimated ride wait times.

Hundreds of thousands of apps compete for iPhone, Droid, Windows or Blackberry space. Many organizations think an app is too expensive for them. They may be right. However, most people think an app is strictly a customized program like the one SeaWorld offers. This type of app often costs $5,000 to $10,000 or more to develop — just for the iPhone or the Droid — so the total price for an app working on all devices could be $20,000 or much more.

However, a much cheaper app is available. It's an icon that's a shortcut to a mobile version of your website. Customizing website pages for smartphones is affordable. The key is to make the site give smartphone users something useful, something they want. My company, for example, customized a website for one of our clients so that the organization's members can use their smartphones to access a membership directory, with just two taps required to email or call a member. The key question to ask, then, is this: "What will people find value in and use?"

Here's another tip. A standard shortcut will just look like a super-miniaturized version of the website page it links to, so customize your shortcut. Typically a miniature version of your website makes for a bad image. That shortcut probably will also include wording (e.g. "home") that may be useless to mobile users. In short, design your icon to support your branding and use wording to provide a meaningful phrase.

Even a space just 45 x 45 pixels, for example, works great for The University of Arizona Athletic Department's ubiquitous red-white-and-blue "A." Below the "A", the default text for this icon could say "Go CATS!" For U of A Wildcat fans, that kind of icon and wording is instantly recognizable and appealing.

The time to join the battle is today. If you were selling home improvement products, would you want to enable smartphone shoppers to search your inventory quickly and easily? The Home Depot does. If you were a private school, would you want prospective parents to easily call the school or access a map, with directions, for a visit? Of course.

In this little-big, local-national, world-wide marketing war, millions of organizations from Arizona Athletics to Zillow are competing for the precious, tiny real estate on a mobile device's home screen.

Be useful and relevant to customers and potential customers and you can win the battle and extend your free lease for a day, or a week or more. Lose relevance and you may get evicted. I deleted the SeaWorld app on our way back to Tucson.

To win, you have to compete. Get to it.

Garmin Versus Droid, and the Cigarette Man Wins

S ome time ago on a vacation, my son and I picked up my nephew for a day trip. Our destination was Isleta Lakes (for fishing), just south of Albuquerque, N.M. Immediately a competition broke out in the car, not about fishing but about technology. Call the competition "Whose Device Provides the Best Driving Directions to Isleta Lakes?"

The announcer on the microphone said, "Ladies and Gentlemen, in the back seat, in black and running without a car adapter, the Garmin, managed by Michael Tedlock, and riding shotgun, managed by my nephew, Dave, the Google Droid."

I immediately agreed to take driving instructions from both contestants. Silence fell upon the cabin of our Toyota Highlander as both managers pressed buttons intently, equally determined to be the first to have his device begin giving directions over the speakers. Meanwhile, I got on I-40 West, which was only minutes away.

Launch time was a virtual tie, as both devices immediately announced that I should "proceed on the indicated route." The Garmin spoke in a man's voice with an Australian accent, the Droid a female voice, with sexy undertones.

Immediately my nephew, a gifted high school teacher and all-around good guy, said, "I tell you what, I'll turn her off and we can just listen to yours, Michael."

The Garmin, with the Droid keeping watch, took us down I-40 to I-25, where we headed south. Isleta Lakes is part of the Isleta Lakes Recreational Area. The Isleta Casino and Resort is plainly visible to the naked eye from I-25, as is the exit.

Inexplicably, the Garmin told us to exit I-25 prior to arriving at the Isleta Pueblo exit. The Droid concurred, so off we went, down into the Rio Grande river valley. The Garmin had us head west for a few miles on a main artery, the Droid concurring. Suddenly, as we passed a large industrial property with a sign that boasted Sheet Metal Works, the Garmin announced "Arriving at Destination." No lakes or water of any kind were in sight.

From the back seat, my son made a puzzled sound and a frantic clicking of keys could be heard. My nephew stepped in, saying, "Well, the Droid says we're not there yet." He turned on her voice. She was sexy. She told us to turn right and shortly thereafter left onto something like "Bakke Road." On we went, at least headed in the direction of the reservation.

However, after several miles of scenic views of towering cottonwoods, a large sign loomed ahead. It read: ROAD ENDS.

So, we reversed course and implemented Plan C — get back on the Interstate and actually take the Isleta Pueblo exit. To our surprise, not a single actual sign by the side of the road suggested the location of the lakes. Finally, we got our directions from a wizened Isleta man who was selling cigarettes out of a single-wide trailer.

The lakes, there for decades, turned out to be lovely, with conveniently located, modern ramadas and brand-new rest rooms with running water. My son almost immediately caught a huge catfish and 20 minutes later I caught one that underscored the monster size of his. My nephew, truly the master fisherman, caught zippo.

Consumers do not spend big bucks on high-tech gadgets expecting them to be as iffy as a fishing trip. We do not set out on a trip and say to ourselves, "Gee, let's try the GPS and see whether we actually get to our destination. This will be fun."

Our goal is not to report, "Well, for several seconds we thought we had a really BIG destination on the hook, but it got away and we realized we were lost."

GPS — in cars, on Garmins and on phones — is still a young technology. When the technology is faulty, we complain loudly. The awful mapping that Apple switched to when the iPhone 5 came out is just another example of how sensitive we are. After all, if we get lost, we can't be sure the Isleta Cigarette Man will be there to show us the way.

Mobile Marketing: What's Over, Now and Tomorrow?

When the iPhone 5 was launched, Apple compared it with the 4S and bragged that the new version was 18 percent thinner and 20 percent lighter and had a bigger display, a faster chip, longer battery life, new Siri features and huge improvements in download times (it's 4G LTE). Projected sales were about 58 million in the first calendar year, and the price of the iPhone 4S dropped, iPhone 4 settling at $0, with a two-year agreement.

Meanwhile, Samsung's Galaxy S3 was already super-popular and powerful, Motorola was bringing out new Droids, and other competitors (Nokia and RIM/BlackBerry) raced desperately to catch up. From now on, it seems that smartphones are going to get better, cheaper and faster at a mind-boggling rate, quickly shrinking the market share of feature (dumb) phones. In 2012, smartphones had already grabbed a 50 percent-plus share of the mobile phone market in the United States.

As a result, mobile marketing continues to grow faster than we can imagine and smartphones have a huge impact on consumers and businesses. This unprecedented speed-to-market makes it hard to sort out which mobile marketing trends are worth pursuing. What is already "Over" (behind us) in mobile marketing? What must we consider right now — what's "Now?" Then, too, what's coming tomorrow? Of course the landscape changes so fast, it's a blurry picture, but here are some still shots.

Over (as in "Toast." "History." "Yesterday.")

Once the focus of so much attention, QR Codes are probably already "Over." Fundamentally, QR Codes fail to offer an adequate value proposition. Consumers must download an app to their phone, use its camera, get focused on a QR code, get a decent photo and then what? Get product information? Not worth it. Get a coupon? Why not deliver the coupon in a simpler, faster way?

Consider Now

No-brainers on the "Now" list are smartphone friendly websites, Facebook/LinkedIn/Etc. check-ins, smartphone aps in general, smartphone-friendly email campaigns and mobile coupons. Smartphone friendly websites, simply put, re-create a site to accommodate the size of the smartphone screen, provide easy navigation, attend to details and offer features such as touch-to-call and touch-to-like functionality. Research shows smartphones will account for over 20 percent of all website traffic in 2012, with that number, as well as e-commerce sales, increasing sharply every year thereafter.

When it comes to building a full, smartphone app, again the value proposition is the key. Bed Bath & Beyond launched an app that was not well received because the app failed to deliver what consumers expected and wanted from Bed Bath & Beyond: coupons. Another national retailer, Michaels, got its app right, delivering coupons right to the smartphone, a real value. Having coupons on a smartphone eliminates coupon cutting and the bother of taking cut coupons with you when shopping.

Two other mobile marketing "Now" options — email marketing and Google Places/Google Plus — merit a quick explanation. Many people first see their email on their smartphone, so email campaigns should be smartphone friendly. Another smart step: use Google Plus to claim a listing in Google Maps so customers on the go can more easily find an office or store.

Now or Tomorrow?

Short Message Service (SMS) or "text messaging" is highly popular, but not commonly used by businesses for marketing for many reasons. For starters, incoming text messaging campaigns must be permission-based. People just do not want to get text messages unless they've asked for them. On the "receiving" end, nonprofits such as the Red Cross have had success offering SMS as an instant and easy way to make a donation. Local nonprofits could use SMS during a major event by making a special appeal, during that event, to donate by text.

Tomorrow?

Predictably enough, mobile marketing is changing so fast — because smartphones are — that it's impossible to name everything in the "Consider Tomorrow" category. The Mobile Marketing Association (MMA) has already established standard ad sizes for ads appearing on smart phones, but most businesses aren't ready to advertise much there — yet.

The Location Based Marketing Association (LBMA) says it's ready to serve an amalgamation of retailers, ad agencies, advertisers, media buyers and Internet service providers who want to buy location-based smartphone ads. Here's an example of location-based marketing. You're headed into Staples when Famous Footwear sends you a message alerting you to the fact that your favorite brand of Nike is on sale just two doors down from where you stand. Consumers are increasingly using all kinds of location-based services, from Foursquare and Neer to Facebook Places and SCVNGR.

Bottom line: Your customers and prospective customers want you to accommodate their smartphones. And they want to receive genuine value if they're going to bother to use their smartphone to interact with your organization.

You don't have to own the latest iPhone, Galaxy or BlackBerry to be smart enough to see the trends. It's ignoring them that would be dumb.

Smart Businesses Serve Smartphone Users

As recently as 2012, if you were to use a smartphone to visit *McDonalds.com*, you would be surprised to discover that McDonald's thinks the action you are most likely to want to take is to "Read the Press Release." If you asked 500 marketing and communication (marcom) experts what the average smartphone user wants to get from a visit to the McDonald's website, the marcom folks would unanimously agree that the opportunity to read a media release would not be the first choice offered to Joe Consumer. The marcomers might say don't offer the release at all.

This startling lapse by McDonald's underscores the jarring facts about the rocket-like rise and fast-growing importance of smartphone use. Today, smartphones outsell computers. In the United States, in 2012 alone, the market share of mobile phone users owning a smartphone rose from under one-third to more than 50 percent.

This rapid rise in adoption is nearly the same for all age groups, from teenagers to senior citizens. Don't kid yourself for a second that smartphones are only popular with young people. It's true people under 35 may use a smartphone as often as a computer to visit websites, but even smartphone owners over 35 frequently visit websites of all types.

Buying habits are changing, too. Best Buy revenues are down and some experts believe that smartphones are a major contributing factor. Amazon may get as much as 20 percent of its revenue from smartphone website visitors who are already spending billions of dollars. By phone. Best Buy's dilemma is that in 2012 industry experts started calling Best Buy "Amazon's showroom."

Consumers who visit a real-world Best Buy store to find the HDTV they want to purchase can use their smartphone to compare Best Buy's price to Amazon's (and to other retailers) and even buy the product from, say, Amazon, while standing in Best Buy. Amazon may not offer installation but often it does offer free shipping.

Local organizations can learn from corporate America's smart moves and dumb mistakes to get and keep a competitive edge in appealing to smartphone website visitors. Here are nine steps any organization can take.

1. Make the navigation and pages smartphone friendly. Use common sense (and maybe skip offering press releases).
2. If you want people to find your physical location,

make the address easy to see and touch. Include city, state and ZIP code so smartphones can map it.

3. Make sure your phone number includes area code and is easy to find and touch.

4. If you have a legitimate Location-Based Service (LBS) to offer, do so. Offer a special discount coupon for lunch or any incentive that motivates mobile users to drive to your location.

5. If you use a QR code, reward mobile users by giving them special information or an offer they can't get any other way. At one big box store, the QR codes posted next to each product led to a website page that simply repeated the information already on display in the store.

6. Because smartphone users are likely to use social media, tie social media in. Offer an opportunity to "Like" you on Facebook or otherwise engage them.

7. Study your website visitor reports for information about smartphone visitors and adjust your site accordingly.

8. Do a cost/benefit analysis before you spend thousands or tens of dollars on a true app for smartphone users. Complete a comprehensive analysis that compares the costs and benefits of a smartphone friendly website to a smartphone app.

9. Be creative. Apple's Siri is not an app, but Apple was and is playful and creative with Siri. Tell Siri, "Beam me up, Scotty," and Siri says, "Sorry, Captain, your TriCorder is in Airplane Mode."

Order Siri to "Tell me a joke" and Siri says pleasantly, "Two iPhones walk into a bar. I forget the rest."

Siri may not tell jokes — yet — but the explosive popularity of smartphones is no joke to organizations that want to stay competitive in the marketplace.

Smartphones and Apps, Search-Engine Ranking, Social Media and Sales

Typically when we think about mobile and its impact on marketing, our first thought is the startling realization that 25 percent or more of the visitors to any website are using their phones to get there. Too often, the site is extremely difficult to use because it's not smartphone friendly. The danger of focusing so much on the need for smartphone-friendly websites is that we may overlook the other major impacts smartphones are having on marketing and sales. This overwhelming list includes: apps, search-engine ranking, email, social media, in-store sales and more.

Initially many predicted that apps would be a key to successful mobile marketing, but the cost to develop an app is prohibitive for many small organizations. For starters, unless you're willing to give up chunks of the market, there's no such thing as one app. At a minimum, you have to invest in an iPhone app, an Android app with support for a variety of phones, and other devices, and then decide whether you can afford apps for BlackBerry and Windows smartphones.

Thankfully, deciding to at least have a smartphone friendly website is a no-brainer — already one-fourth or more of your site visitors are using smartphones and that percentage continues to rise sharply. Another reason for having a smartphone-friendly version of your website is search-engine ranking. Google ranks websites with a smartphone friendly version higher than sites without one.

Here's another communication tool affected by smartphone use: email. Initial studies show people use smartphones to read and send email at least as often as they use a computer to handle email. One impact that smartphone email use has on all sent email is that subject lines need to be shorter. Content also should be tightened up and attachments given extra thought.

Smartphones also impact social media. Facebook has 50-million-plus users who are on Facebook every day, and tens of millions of people use their smartphone to post and get updates daily. Facebook postings, Tweets, Pinterest postings and more are all being looked at, and made, with smartphones. A high percentage of social media users check their phones for updates before they go to sleep at night and first thing in the morning, before they even turn on a computer.

Perhaps the most disruptive impact smartphones are having is on in-store shopping. One commercial realtor reports that commercial real estate agents now call Best Buy stores a show room for Amazon. A staggering percentage of shoppers use their smartphones to comparison-shop while they're in a store. One study suggested that 40 percent of them, when they find the same item online at a lower price, simply buy the item online.

Is it possible that retail stores are headed for the same fate that our nation's newspapers have experienced? Will the weak links in retail in every city go out of business, with the rest left to struggle with an unbalanced and deeply flawed business model as people buy more and more online and have it shipped (for free) right to their homes?

What we can predict is that the mobile market will continue an outward explosion of growth at a pace much faster than most of us can even comprehend. The tablet market has grown so fast that it is already segmenting into submarkets: small tablets and standard-size ones. Smartphones themselves are segmenting a bit the same way, with Samsung bringing out a "small" version of its hugely successful Galaxy 3, the "small" G3 being the size of an iPhone.

Not so long ago, the term "the third screen" referred to smartphones. Screens one and two were television sets and computer monitors. Now it seems we'll have several screens: two sizes of smartphones, two or three sizes of tablets, a handful of sizes related to laptop and desk top computers plus television. Television screen sizes have also segmented more than ever, with screens ranging from 30 inches to 150 inches or larger.

One "fact" most professionals in technology and marketing seem to agree upon is that websites must be "responsive." That is, websites must be programmed to identify the type of device a visitor is using and its screen size, and then deliver up an experience that is appropriate for that user. If your site (and your email) is not responsive, then the price you pay includes lost traffic, lower SEO, less effective email and lost social media opportunities. Any way you add it up, the result is a negative number with a dollar sign in front of it.

Future Shock: The Gadget We Love, Hate and Fear

Consider one writer's comment about smartphones. "There is something compulsive about a smartphone. The gadget-ridden man of our age loves it, loathes it and is afraid of it. But he always treats it with respect." Here's the surprise! Legendary detective novelist Raymond Chandler wrote that in *The Long Goodbye*, published in 1953.

Of course I added the word "smart" to his comment about "the phone," but people today love their smartphones, hate them and fear them. Some of us even experience all three emotions in a five-minute time span.

We are compelled to treat smartphones with respect because their impact on us and our organizations is huge, and growing. It's also true that smartphones caught us, nearly all of us, by surprise. Smartphones represent an unexpected technological tidal wave.

Consider a few facts that may help with this wake-up call. Smartphones are out-selling personal computers, account for over $1 billion in e-commerce per year, and by the time you read this may account for 50 percent of all website visits.

To make sense of these numbers, let's break down smartphone website visits into three categories:

1. *Location-based services (LBS)*
2. *Mobile visits*
3. *Routine visits*

Droid and Apple smartphones give you turn-by-turn directions and deliver advertising to you based on your location. Apple's iPhone already offers over thousands of LBS apps. A Target app directs you, when you're in a Target store, right to the vacuum cleaner bags you're looking for. That's LBS. You don't need a sales clerk to help you find a product in the store. Your phone will handle that for you — and for Target.

Now hit the pause button on smartphone website visits for a moment to consider a myth about apps. A businessman I know once displayed his temporary ignorance by pronouncing that "apps are games."

"My BlackBerry," he said without irony, "is for business." A few months later he switched to an iPhone.

Still, it's tempting to try to ignore business-oriented apps and dismiss LBS as only important for consumer-driven retailers. If you offer no specials, won't people just use their computers to visit your website? No.

We don't wear our computers. People are compulsive about smartphones because they always have them on and they can do so much. While people are waiting for a meeting to start, for a child to get out of school or for any other reason, smartphone users can read their email, check headlines, text people, log into Facebook to see what's up with their friends, and more.

In short, people on the move may choose to visit your website with their smartphone just to save a few minutes. When they do, they expect and need to see a mobile version of your site, one that makes it easy to call you, locate you on a map, navigate pages and more. The good news is that we can use website programming to recognize a smartphone when it visits and serve up a site uniquely suited to that visitor.

It's vital to expand your thinking about a mobile version of your website beyond people who are just "stuck" without a computer. Smartphones provide a fast and convenient way to visit websites. This past Friday night, my wife and I sat at home and wondered what movies were in town. I turned on my computer to check listings; she grabbed a smartphone. By the time I was typing in "movies.com" in my browser's address bar , my wife was already telling me what movies were at our favorite theater. Even I was surprised. So speed is another factor in smartphones getting used before a computer is.

We're unprepared for what smartphones mean to organizations because technology is developing faster today than our ability to comprehend its impact. Alvin Toffler described our situation as "...the shattering stress and disorientation that we induce in individuals by subjecting them to too much change in too short a time." Toffler called it Future Shock, the title to his book published in 1965.

If you are a leader in your organization you should already own a smartphone. If you don't, you can't understand how most adults think about mobile. If you are in marketing or sales, you can't fully understand all your customers unless you're using a smartphone yourself. You can deal with the love-hate relationship and with your fear, but you can't deal with your own ignorance. You don't know what you don't know.

In *The Long Goodbye,* detective Philip Marlowe is relentless in his pursuit of the truth about how a woman died and whether the friend he drove to Mexico was in fact the killer. Now we have to be relentless and make sure we know at least as much as our smartest smartphone customer.

The future shock we should fear the most is that we won't learn what we need to know about customers until it's too late to stay in the game.

Chapter 10

Search Engine Optimization (SEO)

Maybe the Risk of Getting Blacklisted by Google Is Worth It

A few years ago residential real estate appraisers found themselves in an extremely difficult situation. Not only had a huge drop in home sales caused a huge drop in their business, but the government had stepped in and completely changed the rules by which appraisers got their business.

Before the banking/mortgage/housing crash, appraisers depended on their reputation and personal relationships to get work. After the crash, simply put, the government required that appraisers be selected (almost) anonymously.

One appraiser asked us for help with a crazy selection system in use by a huge bank. It was wacko! Preapproved appraisers had to log into a website to see if any appraisal work was available. If so, it was first accepted, first selected. The problem our appraiser faced was that no matter how many hours he spent logged in every day, every single appraisal was taken in a split second.

We quickly realized the problem wasn't our client's bathroom break on Tuesday at 10:32 a.m. That's not why an appraisal request appearing at 10:33 was snapped up by another appraiser with greater endurance.

Clearly, some appraisers had found a way to cheat the system. We learned they had purchased illicit software from an Asian company that automatically logged them in every half-second of every hour of every day and grabbed any work offered. Ca-ching!

The terms of the appraiser's agreement with the bank clearly stated that this approach could result in the appraiser getting blacklisted, but I talked to appraisers in many states who said they did it anyway. An appraiser in California remarked, "Yeah, we got blacklisted, but it was great while it lasted. We had a ton of work."

Today, business owners face a similar paradox with Google and their search engine ranking. Get onto Page 1 of organic search results, especially in the top three, and your business will get the leads needed to stay in business.

However, getting there can seem like Mission Impossible. The book, *Search Engine Optimization for Dummies*, is more than 500 pages long. And even that book is a simplified look at the complex subject of Search Engine Optimization (SEO).

At my company, we've studied books, read articles, attended seminars, subscribed to newsletters, used dozens of techniques and here's the truth. Legitimate SEO is hard work and never produces instant gratification.

Cheating, on the other hand, is effective.

The fast track to Page 1 search engine results is to scam Google by taking the exact approach that Google warns against — by buying into a sophisticated link-farm system. A link farm is a system of hundreds or thousands of websites, usually nearly all of them fake, on which an organization is listed and a link to its website is provided.

Google wants us to think that its method of search-engine ranking is so incredibly well conceived and pro-grammed that links on fake sites won't have an impact, but in fact, they do have a huge impact. Google threatens to punish cheaters and has in fact done so. Not all that long ago, J.C. Penney was a notable example.

Today, small organizations try the link-farm system for a simple reason — they want to stay in business and be profitable. The plan goes like this, "Get on Page 1 of search engine results, make money now, stay in business, and maybe Google will never notice." The risk of getting penalized by Google seems small compared to the ben-efit of making money today.

Buying into a link-farm system is no guarantee of a Page 1 listing, but sometimes it works. What will happen, in the long run, to the organizations that are cheating on Google?

Consider the fate of residential real estate apprais-ers who gamed the bank's system. It took months for the bank to realize some appraisers were gaming they sys-tem. Appraisers who played by the rules got hurt.

One appraiser who was blacklisted in the end told

me, "We get nothing now. Basically, I'm out of business." Even so, he added, he was glad he'd grabbed the business when he had the chance. Doing so kept him profitable while his competitors went under.

What would you have done if you were the appraiser? We already know what some business do — they break the rules and hope they don't get caught. Sometimes the possibility of getting blacklisted by Google is worth the risk if it saves a business this week, this month, this year.

Help Yourself with Search Engine Optimization (SEO)

When business owners or marketing people think about Search Engine Optimization (SEO), it's tempting for them to conclude that nearly all SEO work must be done by website programmers. That's not true. Here are several steps a business owner or marketing and communication (marcom) person can take to improve your Search Engine Ranking. Some steps do require assistance from a website developer, but none are mysterious or super-high tech.

First, make sure your domain name registration expires at least 10 years from now. Search engines monitor domain name expiration dates. Which website seems most likely to you to stick around, a site with a domain expiring this year or a site with a domain expiring in 2025?

Next, assess, or reassess, the keywords you think people are using to try to find your site and your competitors' sites. Look at your visitor logs. Survey your customers formally or informally. Check trade publications for articles that include keyword information. Then identify the keywords you want to focus on in your SEO strategy.

Select realistic keywords for SEO. Occasionally, when we ask a client for keywords, the answer might be "interior and exterior residential, commercial and industrial painting in City X." It's difficult to rank high in many broad categories, so be specific. For example, maybe "Tucson exterior house painting" are the keywords.

Once you've identified your keywords, check your current website for its keyword use. Your website developer can run a word count on each page as well as one for your entire site. You may be surprised to learn you're using the word "Tucson" only five times in your entire site when you could easily improve SEO by working it in 30 times, without "stuffing."

"Stuffing" means using keywords over and over again in an unnatural way. Avoid stuffing. Search engine algorithms approximate the way human beings read your site content. If you repeat keywords too often (in this article, for example, "SEO" certainly gets repeated a lot), your content can sound awkward or forced. As a result, you may get penalized. One SEO guideline suggests using your keywords no more than four to six times per 350 words of content.

Even so, we know one site that is definitely guilty of stuffing keywords, but it still has great SEO — Google ranks it first in search results. Therefore, you may want to push the keyword envelope and move close to real stuffing.

Even so, avoid duplicating content, either on pages within your own site or by borrowing content from other sites. Copying the content of other sites (even when you aren't violating copyright) is a bad idea from an SEO standpoint. Instead, write fresh content and don't share it with other sites.

You can also write actual words that don't appear on the site, but do appear in search engine results. These words are placed in the page title and description for each page. Some website developers don't ask their clients for this content and don't provide any content themselves. We've seen entire websites with the same page title, "XYZ Corporation of Tucson, Arizona," and the same description, "The XYZ Corp makes great widgets," on every page. Make each page title and description different and relevant to its content — and don't forget to focus on your keywords.

Here's another tip. When you are writing a link — to one of your own pages especially — don't just write "click here" and make those two words the link. "Click here" has no contextual power. Instead, write "Learn more about SEO," for example, and use that phrase as both the link and the description of the content a person will see if they click on that link.

Lastly (at least for now), spend extra time considering headlines and images. Search engines like headlines, subheads and sub-subheads, so pay attention to their wording. Headlines have more SEO power than body copy, so make them count. Then review all the images on your site and ask two questions: "How are the images named?" and "What does the alt tag say?" The names and alt tags of images represent two more keyword opportunities.

Some websites use images with file names of zero keyword value. Starwood Hotels, for example, starwoodhotels.com, has great photos of the Westin La Paloma, but here's a sample file name: *wes1001gr.14735_ub.jpg*.

Instead, the file name should be descriptive. "Tucson pool-side views at La Paloma" is just one example. The alt tag refers to "alternative text" attached to an image. Make alt tags count by providing descriptive wording.

Now, a quiz. What keywords are used most in this column? The answer: "keyword" (used at least 20 times). That's not great. However, SEO also does well, being used 13 times. Or should SEO be used 14 times? Did the acronym "SEO" seem to you to be used too much? The keyword part of SEO is up to the marketing people who are keyword specialists, not to the programmers.

Chapter 11

Security

Your Email Is Gone, Smartphone Is Dead and You Lost $12,345

Suppose you get up one morning and discover all the photos and music you had stored online are gone. Everyone in your email address book has received the first in what will be a series of racist and homophobic emails from you. In addition, "you" will be sending out email bashing small-business owners as greedy scumbags.

Your day gets much worse. Your smartphone is dead. Your email account and archives are gone. And $12,345.55 has been stolen from your business banking accounts.

Last night, you got hacked.

According to Mat Honan, in "Hacked," in *Wired* (December 2012), a competent hacker can probably get into your accounts in less than an hour. If you own a small business, you are more likely to be a target.

Here's a short explanation of how people get hacked. The primary weakness is the username — for most people their email address. A hacker expects a person's email address to be the username. Email addresses are easy to get. The second weakness is the use of easy, predictable passwords. That includes a popular choice of "password" for the password! Really. A third weakness of your online accounts is that when major corporations are hacked, their lists of usernames and passwords are often dumped on the Internet for any other hacker to use. Tens of millions of accounts posted up with a "Come and Get It!" invitation on top.

Suppose you haven't already been exposed to hacking caused when a major provider of services had a security breach. Hackers still have several powerful tools with which to ruin your life. One, visit the wrong website and malware (software that sends your usernames and passwords to a hacker) may be installed on your computer without your knowing it. Two, email phishing schemes may con you into entering your username and password on a bogus site. Three, guessing the right security answers may enable a hacker to reset your password. Four, a "brute force" attack may work against even a moderately strong password. Five, hackers succeed with other approaches, including the use of a call to or chat session with tech support to report "a forgotten password" and have the password reset over the phone.

A hacker's highest priority will probably be your email account. Access to your email account provides buckets of info about you plus a quick way to reset passwords and access other accounts.

If you're using IMAP (your email is in the cloud),

then the hacker who gets into your account can search through saved emails and use phrases such as "bank account," and "password" to find accounts, usernames and passwords.

Concerned yet? You should be. Here are steps you can take to protect yourself.

Step 1: Don't click on email links and then enter your username/password information.

Step 2: Don't use the same username and password for every account!

Step 3: When you can, choose a username that is NOT your email address. That makes it vastly harder to guess.

Step 4: Use a random series of letters and numbers for each password. As of today, a brute force attack will fail against a strong, 16-character password.

Step 5: Pay special attention to the usernames and passwords for vital accounts such as email, banking and online stores. It can really cost you if such accounts are hacked; having other accounts hacked may cause you major embarrassment and/or inconvenience but not cash.

Step 6: Don't provide typical answers to security questions. Hackers can probably find your mother's maiden name on the Internet, so make up nonsensical answers to common security questions. If the question is, "What elementary school did you attend?" your answer could be "My Big Fat Aunt Winona" (my apologies to Aunt Winonas everywhere).

Step 7: Do not make a list of all your usernames and passwords and save them to your smartphone or laptop. If you do and either gets stolen or hacked, you're screwed. Instead, print out a username and password list without saving the file.

Step 8: Don't share your usernames/passwords via email. Email is not secure unless you pay for secure email.

Now if you've been paying attention, you may be thinking (correctly), "Great, but if my email does get hacked somehow, the hacker can just reset all those fancy usernames and passwords I created." That's right.

Therefore (ideally), take Step 9: Create a separate, private email address you use only to manage your online accounts. Do not share that email address with anyone.

If you follow these steps, you'll be better protected than 99 percent of Internet users. Better to be in the 1 percent and super hard to hack than among the 99 percent.

Among the 99 percent, some are going to wake up tomorrow with a dead or missing smartphone and a big hole in their bank account. You don't have to risk being one of them.

The Power of Technology, Shrinkage and Rude Ralph

To get through graduate school, I worked retail as a department manager at the Harvard-MIT Cooperative Bookstore (The Coop). The Coop then was really a small department store with five locations.

As a department manager at one of them, I saw management puzzling over a particular delivery driver whom we'll call Ralph. He was rude, loud, foul-mouthed and always rushing around like a guy on amphetamines, racing up to the loading dock and then slamming on the brakes so that they screamed in protest.

Oddly, Ralph was not a particularly productive delivery man. One day I saw the No. 2 guy at The Coop, clipboard and stopwatch in hand, riding Ralph's route with him. Clearly management suspected Ralph of some kind of wrongdoing greater than rudeness and bad language.

Two years later I had my graduate degree and my second teaching job, this one in Boston in The Coop territory at the business school. One day as I walked down Mass Ave. for lunch, I heard a familiar sound behind me — a certain vehicle braking abruptly to a stop. I turned around. Sure enough, it was Ralph. But he wasn't pulling up to a loading dock. He was double-parking right there on Mass Ave. Then he rushed to the rear of the Coop truck, threw open the doors, grabbed a couple of bulging shopping bags, walked quickly to the rear of a car parked on the street, popped its trunk, inserted the bags, then drove off at maybe 20 over the speed limit. Clearly he was stealing merchandise. Always had been.

Today's technology could catch Ralph when, evidently, top management hadn't been able to nail him back two years earlier.

Consider today's options. Today, surveillance cameras are cheap and easy to use. At store loading docks, in warehouses, even on the back of vehicles. They could have observed and recorded everything Ralph put in — and took out of — that truck, including those bags. Surveillance video is now digital, which makes searches faster and easier.

According to Bill Gaither, General Manager of Accura Systems of Tucson, "The most common use of video cameras is to reduce shrinkage [theft] — from outside, or inside, the organization." Video cameras do have limited focus areas, so having too few cameras can leave a business with less-than-clear images.

Then, too, Gaither admits that he himself has seen video in which the thieves wore masks. In one case, the video recorded the masked thief stealing the video camera!

Security technology goes far beyond video. Other common tools are access badges, panic buttons, money clip alarms and security alarms. Access badges restrict access to one or more areas on a building. Panic buttons silently call 911. Money clip alarms attach to the last $20 bill in the drawer, which cashiers are told not to remove. In a robbery, the robber is gladly given all the $20's. When that last bill is yanked out of the drawer, the clip activates a silent "robbery in progress" call.

Security systems are surprisingly versatile, too. Supervised opens and closes record the exact time at which a given employee — or vendor — turned the system off, or on. So the system reports exactly how long the cleaning crew was in the office, or whether the business really had Mary Smith in by 7:55 a.m. Sophisticated systems can even be set to send you an email when the system is turned on, or off, or turn on the video cameras the instant the alarm is set off, saving gigabytes of hard drive space.

Of course people must support the technology. In Tucson, the police will not necessarily answer a call when a security system alarm is set off. At least two security companies in Tucson can be paid to send one of their patrolling security guards to a site immediately if an alarm is triggered. That response can make a police visit guaranteed if criminal activity is under way, or unnecessary if the alarm is false. Business owners who don't like being awakened in the middle of the night over a false alarm often pay for this option.

Lastly, delivery vehicles (remember Ralph?) can be outfitted with a GPS (global positioning satellite) device that reports to the business owner, via a password-protected website, amazing details about each vehicle's travels. Discreet Wireless sells a system that Sparkle Cleaners in Tucson uses. Heath Bolin, president of Sparkle Cleaners, says, "One of the largest liabilities a business owner can have are its drivers out in company vehicles."

Discreet Wireless's GPS system reports a vehicle's current location, maximum speed, time between stops, location and length of each stop, and (depending on configuration) the exact route (a bread crumb trail) the driver took. Really. The reports are easy to generate and beat the heck out of grabbing a clipboard and riding with a driver.

Thinking back, clearly Ralph parked his car on Mass Ave. to save time because Mass Ave. was on the route between stores. So Discreet Wireless's system would not have shown Ralph off course or stopping long enough on Mass Ave. to raise suspicions. But Ralph would've been caught speeding.

Back then, I shook my head and reflected that it was easier to teach about shrinkage than actually stop it. Today, a video camera on the truck — or probably just cameras at each loading dock — would nail Ralph. After all, his only mask was one of rushing and rudeness. Cameras would have seen right through that. Instead, Ralph appears on candid camera only in this column.

Costly Security (Antivirus) Software Mistakes to Avoid

Uninformed and inattentive people can make extremely expensive mistakes related to computer security software. This software is sometimes often inaccurately called "antivirus software." These mistakes can cost individuals, and the organizations they work for, serious time and money. Here are some guidelines for what to do right, and what to avoid doing wrong.

Keep Your Security Software Current

Norton 360 and McAfee are just two of more than a dozen standard brands of security software. These programs rely on a "subscription process" that constantly updates a PC with new protection against new threats. Typical subscriptions are for a year, but sometimes a brand-new computer will provide 30 or 90 days of "free" use first. Unfortunately, some people let subscriptions lapse. The subscription must be renewed and kept current.

Keep a List and Check It Twice

When more than one PC is involved, at home and/or at the office, keep a list of the name of the machine, the security software installed and its expiration date. Don't assume each family member or employee is keeping the software current. Check. We've heard employees say, when asked about an alert that a subscription has expired, "Oh, I just close that window so I can use my machine."

Know Who Is Using Your PCs

One office we know of allowed the full-charge bookkeeper, a single mother, to complete some of her work on the weekend, when her son was not in school. What the company didn't know was that the bookkeeper kept her son busy, while she worked, by logging him onto a secretary's PC. The boy then downloaded and played dozens of video games on that machine. Some of those games infected the machine with viruses.

Establish a policy about who can use computers. Keep separate passwords for different users on different machines. Make sure automated scans are set up to detect problems.

Computers That Are Turned Off Can't Update Themselves

Occasionally a well-meaning Information Technology (IT) person will set PCs to automatically update virus software and run security scans when an organization is closed for business. This approach makes sense in that scans require a high percent of computing resources to complete. No one likes using a PC that's running a complete scan. In addition, by using a machine during scan-and-backup, the user takes a risk that the scan will not complete.

A problem can arise, however, when updates and/or scans are set to run at, say, midnight. Suppose the employee isn't told about the backup setting or just forgets about backup and just shuts the machine off at 5 p.m. every night. The result is a machine that never gets viruses updated and is therefore vulnerable.

Tell Employees to Ask about Registration Renewal and Be Careful

Computer fraud is at an all time high. Here's one of the latest gambits. Your employee or family member gets an email or on-screen message that says, "Warning! Your computer has a virus. Your security is at risk. Click here to run a scan now." A variation is, "Your antivirus software has expired. Click here to renew now."

The scam is that the link for the free scan or for the subscription renewal is to a criminal's website. The logos and wording may look right, but the email is spam and the site is a fraud. In one case, a client actually "renewed" his security software at a fraudulent site and paid for it by credit card! The same day, his card was used fraudulently, and meanwhile, his computer had installed not security software, but a deadly program instead.

If you are not sure whether you actually need to renew your security software, take these steps. One, do not click on a link sent to you in an email. Two, consult your list and determine what software you have installed on your machine and when it expires. Three, open that software program on your machine, if it is set to expire soon, and click on the link, in the genuine program, to renew. Or go to the website for that program to renew. You may also call the 800 number for the company in question and get tech support there.

Lastly, if you suddenly realize you may have connected to a fake or fraudulent website, **do not just close your browser** (Internet Explorer, Safari or Firefox). Instead, **turn your PC off.** Then get expert help immediately. If you feel you must turn your PC back on, disconnect it from the Internet and/or your computer network first. That way, at least your PC won't try to connect to the fake site or infect other machines on the network.

Computer fraud costs are estimated to be as high as $500 million a year. Take the right steps and you can make sure that $0 came from your pocket.

Chapter 12

SMS / Text Messaging

SMS 101: What You Need to Know about Text Messaging

Pop Quiz. Which best describes your view of text messaging and your smartphone?

A.   Great for telling my spouse what to get at the grocery store.
B.   How else am I supposed to reach my teenager?
C.   I've seriously considered its use for communication and/or marketing within my organization.
D.   Text messaging has significantly increased our business in the last six months.

If you answered D, email me. With your permission, I'll share your success story in whatever way I can.

Everyone else should find some value here in this SMS (Short Message Service) primer.

First, Understand the Technology

Understanding how Text Messaging works and can

be used as a tool is pretty easy. Your challenge will be whether you can implement an effective communication strategy, using SMS.

Two Sentences on History

What was originally called SMS, we now just call texting, or text messaging. Initially text messages were short and pure text. Now text messaging options include the sending of images and video. For marketing and communication, that option is a game-changer.

A Simple Outbound SMS Example

An outbound text messaging campaign uses a list much the same way an email campaign does. Important note: for email campaigns, opt-in is strongly recommended. For SMS, "intentional" opt-in is a Federal Communication Commission (FCC) requirement.

Here's a simple, but flawed, example. A man creates a list of more than 100 cellphone numbers and uses the list to send one text message, with a photo attached. Every person receiving the message is pleased, especially to receive the attachment. The man's personal use of SMS is different from the legal requirements that apply to an organization.

Two Options for Organizations to Send and Receive

Generally, organizations that want to implement SMS choose between two basic options for text messages.

One is short code. The other is an enabled phone number, also known as long code.

You've seen short codes used on television. A five or six digit code is used when the call to action is, for example, "Text 49494 to Vote for Beverly!" Short code numbers must be purchased from the Common Short Code Administration (CSCA).

Short code use requires additional steps, which include planning and probably paying a specialist to help with implementation. Short codes are for lease only. Typically, short codes are for high volumes and big budgets.

Long Codes, aka Enabled Phone Numbers

SMS-enabled phone numbers (long codes) are much more affordable than short codes. Long codes can be used for both sending and receiving text messages. The code is "long" because the number is 10 digits — a phone number including area code. These numbers are inexpensive, costing as little as $5 per month plus per-message fees.

Needed Volume Makes a Big Difference

Short codes can send more than 2,000 messages per minute; long codes may be limited to 10 per minute. Suppose that an organization decided getting 600 messages out per hour was too slow. The organization could use multiple phone numbers (long codes) to speed up the process.

There are some other differences between short

codes and phone numbers, so study up if your text messaging becomes a critical strategy for you.

Thinking of Using a Smart Phone? Be Careful and Know the Law

You might already be thinking that a simple, cheap SMS approach to list management and decent volume would be to simply use a smartphone for text messaging. Lists can be managed on a MAC or PC, after all, and then downloaded to the phone. Apps exist that help with the process. Even at only 10 messages out per minute, or 600 per hour, a dedicated smart phone could still do 14,800 text messages a day.

Here's the catch. The FCC requires an organization to document outbound SMS use as intentional opt-in. In other words, if you're going to long code text people on behalf of an organization, you have to have proof, separate from email opt-in, that those people opted-into your text messaging. You also have to show there's an opt-out option.

Marketing Strategy: More Difficult Than Technology and the Law

At a minimum, the most successful outbound SMS strategies will include these fundamental elements:

- Documented, intentional opt-in with an opt-out option.
- A relationship with trust.

- An important message.
- Urgency (otherwise email would do).
- The delivery of value to the recipient.

These fundamental elements merit a separate discussion, so a simple example here of personal use (not organizational) will have to suffice. This example could only work for an organization if the intentional opt-in/opt-out requirement was met.

A man, Noah, wanted family and friends (a relationship with trust) to know immediately (urgency) when Noah's wife gave birth to their baby. The news, and the photo of the new-born son as an attachment, had great value to this target audience, so they are thrilled to get the text. I know because, I received the text and the photo. What a cute baby!

You may never quite match this perfect blend of audience, urgency, relationship and value received. Even so, at least you know the long and short (code) of it.

Chapter 13

Social Media

Legal Words of Caution about Social Media

A quiet panic had settled over the room. The speaker, a lawyer, was giving dozens of ad agency executives a lecture on a seemingly-obscure yet chillingly-familiar topic: Social Media and Truth-in-Advertising. The part that had people scared to death had to do with disclosure and accountability.

A woman in the back timidly asked, "You mean we have to make sure a blogger or reviewer who got compensated somehow discloses that?"

The room seated more than 400 people and you could hear a pin drop.

"Absolutely," the lawyer said. "The FTC is very particular about this aspect of Truth-in-Advertising."

A woman sitting next to me turned pale.

For once, I had an edge over the enormously successful ad agency executives in the room. I am married to a lawyer. For years she has facilitated my attending seminars on Internet-related legal issues given by lawyers for lawyers. I'm not a lawyer, but I've attended legal presentations about domain names, meta tags, trademark infringement, intellectual property rights and more.

Here's a fundamental fact some ad agency people and many business people forget. It's a principle you need to know. Despite myths to the contrary, the Internet does not supersede or erase the law. For example, the FTC's Truth-in-Advertising rules apply to social media. Here are two examples.

First, consider Twitter and tweets. Suppose a local "celebrity," the head coach of the university's men's basketball team, is paid to endorse a product or service. Suppose the coach Tweets about the product. Each Tweet, within its 160 characters, must disclose that he is a paid endorser. Really.

The FTC's website clearly states the fundamental principle that applies: "The Endorsement Guides also state that if there is a connection between the endorser and the marketer of a product that would affect how people evaluate the endorsement, it should be disclosed." Suppose a coach, a celebrity or anyone else is paid to Tweet positively about a product or service. That payment does affect how people evaluate the endorsement. Therefore, the Tweet itself must disclose the fact that compensation is involved.

A second important example involves reviews and comments such as those posted on Google, Yelp!, Best Buy and countless other sites. Assuming a person's comment and/or review is positive, that means the person has provided an endorsement. The FTC says, "... the Endorsement Guides let endorsers know that they shouldn't talk about their experience with a product if they haven't tried it, or make claims about a product that would require proof they don't have."

Countless times companies have paid people to write positive reviews of products those people have never used. One such attempted payment took place through an Amazon service called Mechanical Turk. Mechanical Turk enables a person or organization to pay people to do what could be called online piecework.

What seems like a well-documented example of misuse involves an executive at Belkin. As you may know, Belkin is a seller of computer accessories. The Belkin executive allegedly used Mechanical Turk to solicit people to write fake product reviews. He didn't ask people to even try the products. He just wanted five-star, rave reviews. Google the right phrase and you can probably find a screen capture of his notorious solicitation on Mechanical Turk.

The FTC does not have a big enough staff to catch every lawbreaker, but people do get caught — and fined. At the conference I attended, the lawyer gave examples of companies, big and small, who had been fined by the FTC.

That day at the conference, the lawyer made one last point that particularly cast a pall on the room. He said the "advertiser" was responsible for the conduct of reviewers or bloggers or anyone the advertiser compensates.

Suppose an organization pays a blogger to blog about a product. What if the blogger blogs but fails to disclose the payment? The FTC is not interested in trying, itself, to hold every blogger, commenter, reviewer or others accountable. Instead, the FTC holds the organization accountable for appropriate disclosure. Because an advertising agency is acting as the agent, for an organization, the FTC also holds the ad agency accountable. This accountability extends into every single social media effort.

Protect yourself and your organization. Keep in mind this very first sentence from the FTC's home page: When consumers see or hear an advertisement, whether it's on the Internet, radio or television, or anywhere else, federal law says that ad must be truthful, not misleading, and, when appropriate, backed by scientific evidence.

Don't let the word "ad" let you think reviews, comments and blogs are exempt. They are not. If your organization retains an ad agency for this work, make sure the agency provides proof that disclosure is being included.

One last tip: Be cautious the next time you see 47 reviews all giving a product a five-star rating. Real people just don't rate anything that uniformly, unless they are paid to do so.

Chapter 14

Tech Support

13 Steps to Success in Calling Tech Support

Chances are, you hate calling tech support. Most of us do. Even so, at times you'll have to call anyway. When you do, these 13 steps to success can make a huge difference.

Step 1: Google It.

Start with Google. Google the problem you're having. If you get an actual error message on your smartphone, tabletop, laptop, desktop monitor, big screen TV, etc., then Google the exact error message. You may luck out and learn how to fix the problem yourself. What a great outcome! At a minimum, you will probably learn something that will help you with your call to tech support.

Step 2: Call during Eastern Time Business Hours.

All of the tech consultants and geeks I have talked

with agree: you are more likely to get good tech support if you call during standard business hours. Call on a night or weekend, and the odds go up that you'll be frustrated.

Step 3: Be Positive.

Do you hate calling tech support so much that before you pick up the phone you're feeling cranky? That's bad. You've already taken a big step toward failure. No tech support person wants to talk to someone who's in a bad mood. So be positive.

Step 4: Be Prepared.

You'll almost certainly have to provide an account number or username, plus a password, just to get started. You may need license numbers, version numbers and more. Make sure you are ready. If you don't know what you need, call and find out. Then call back when you've got all needed info on hand.

You may be thinking, "Can't I just kind of stumble through the 'account authentication' process and/or any other requirements?" You can, but it's a bad approach. Don't use up your tech person's patience on something as stupid as your failure to look up your account information. Tech support people don't have endless patience any more than you do.

Step 5: Be Charming.

The "be charming" commandment can be a truly tough assignment. Try anyway. For starters, think of the tech-support person who takes your call as an actual human being. "See" that person as having actual feelings and a life outside of work. Imagine that person stuck in a cubicle, sitting on an uncomfortable chair, wearing an uncomfortable headset or ear piece, and struggling to hear you well because of the chatter in the room and a poor phone connection.

Next, understand that tech support people are often frustrated with a limited menu of resources they can use to identify and solve your problem.

Lastly, assume a genuine idiot talked to your tech support person just before your call. Assume that last caller was both stupid and rude.

Step 6: Don't Poke the Bear.

If your tech support guy tells you, over a bad phone connection, in a heavy Asian-accent, that his name is "Jack," do not say, "Yeah right. What's your real name?"

Instead, say, "Thanks, Jack, for taking my call. How is your day going?"

You may get stunned silence. "Jack" may have never had a caller so considerate and friendly. Listen for the reaction you get. Then continue to be friendly and courteous.

Step 7: Don't Position Yourself as Stupid.

If you must describe your own reaction to the situation, do not say, "I'm a complete idiot when it comes to…" Jack needs all the confidence he can muster. Learning that he's talking to an idiot is not going to help him — or you — succeed.

Step 8: Slowly and Carefully Explain the Problem.

Speak up. Enunciate. Consider the fact that English may not be Jack's native language. That means he has to translate whatever it is you're saying into his own language. If he finds a solution, he has to explain it all to you.

Step 9: Hang Up and Try Again as Needed.

In the first few minutes, you must answer a crucial question for yourself. You have to decide whether "Jack" is competent. If he solves your problem on the first try, congratulations! If he puts you on hold for several minutes, then comes back only to promptly put you on hold again, that's a bad sign.

If you think Jack's incompetent, here is what you can say (interrupt him if necessary). "Jack, I am so sorry. I have to go now. Thank you for your effort. I will have to call back later."

Then hang up. Redial tech support, and try the next guy out. When you call back, you will almost always get a different tech support person. Make no reference to your first call. Just start over again.

Not all tech support people are equally capable. If you get an idiot, get rid of him or her. If you get a genius, ask whether, in the future, there's any way to direct dial to his or her desk for tech support.

Step 10: Take Notes and Keep a Record.

If at all possible, take notes during the call. You may need them later. If you get an email with a "ticket number," save it. If your problem remains unresolved, use any tickets to help you get a supervisor to help.

Step 11: Implement the Solution while Tech Support Waits.

If at all possible, get tech support to stay on the line with you until you are sure your problem is solved. You may have to reboot your computer, or reload a program and open it, or take other steps. Do it.

Step 12: If You Must, Ask for a Manager or Supervisor.

It's always OK, if your tech support person has truly failed you, to ask for a supervisor. Supervisors are sometimes more highly experienced and successful — that's how they got promoted to supervisor. One downside is that you may have to wait until they get off the call they're on.

Now, here are two other points to always keep in mind. One, you may want to ask, "How is that supposed to help?" If your tech support person knows the answer, that's a good sign. Two, refuse to take a step that seems stupid or crazy, such as, "Now we're going to re-format your hard drive!" See Nine — Hang Up and Try Again.

Finally, a baker's dozen.

Step 13: Get Someone Else to Call Tech Support.

For starters, recognize that your own time is valuable. Don't waste it. Worse yet, if just reading these tips for tech support success makes you physically ill, stop tormenting yourself. Get some help on your end, whether it's an enlightened teenager or a high-powered technical consultant. Your time — and sanity — are valuable. Proceed accordingly.

Outsourcing Problems Run Deeper than Names, Accents or Phone Connections

A man I know in the aircraft industry tells this story about calling tech support. He quickly detects that his tech support person is Asian, working somewhere in Asia. The tech support guy has said his name is an unlikely one — "Tony."

My friend has read Thomas L. Friedman's book *The World Is Flat*. He understands instantly that "Tony" has been given this American name to make his callers feel more comfortable than they would with, for example, "Nyun."

So my friend playfully says, early on in their conversation, "Hey Tony, what's your last name?"

Tony says, "Just a minute. I put you on hold."

When Tony comes back on the line, he says, "They haven't given me one yet."

You don't have to be a geek to have had problems with tech support that's been outsourced to Asia. Sometimes the problem is that the actual connection's not great, and that makes it hard to hear. Sometimes the problem is that your tech support person doesn't speak English well enough to deal with your idiomatic phrases. Sometimes the problem is that "Tony's" English is just difficult to understand.

Anymore, most Americans are hugely relieved when we recognize that the voice on the other end is native to the United States. No matter how thick the Southern accent or how twangy the Texas drawl, that voice is, to us, a relief. Lovely, even.

The mistake companies large and small are now making in outsourcing goes beyond problems with a foreign accent we struggle to comprehend or a bad phone connection.

Consider this experience. A company, we did business with for several years was acquired by (I know because I asked) a company in Malaysia. I'll call this company "Formerly," as in "formerly a United States of America company."

Unaware of this change in ownership, we purchased technical services based on specifications listed on Formerly's website. However, as we implemented these services, we discovered that important elements listed in the specs were no longer offered. This fact surprised even our American sales contact, who agreed that this meant that Formerly's site was not simply misleading but outright wrong. He was apologetic but powerless to get us a refund. We had to call billing, in Malaysia.

The calls to Malaysian billing went badly. The customer service agents and even the managers I talked to did not understand these simple points. One, we were long-time customers of Formerly. Two, we paid for services based on information that was still, weeks later, on Formerly's web site. Three, because the services promised, in writing, were not available the package was useless. We wanted a refund.

Even a Malaysian supervisor of managers just said, "No refund." The facts meant nothing. There was no comprehension. No apology. Just, "No refund."

In trying to have a discussion with her, I remained polite and calm and used phrases such as "truth in advertising." I sent her the link to her own site.

"No refund."

The U.S.A. sales person did not answer my call, but I suddenly had an idea. I Googled "Formerly" and learned — imagine my surprise! — Formerly still belonged to the Better Business Bureau of Atlanta. At the BBB's site, I completed a complaint form. Weeks later, I got a call from a Formerly Vice President of Sales, an American in Atlanta. He was genuinely embarrassed about the BBB complaint and apologetic. We received a 100 percent refund that same day.

Businesses succeed in the United States based on a code that's based on a culture. For example, we value long-term, ongoing customers. Sure, too many American business people screw customers over, know it and don't care. The problem with my Malaysian supervisor, however, was that she clearly did not know the code. She wasn't scripted to know our culture any more than Tony was scripted to have a last name. And even on a support call, that's the most important script there is to know.

Retaining the Wrong Computer Consultant Can Be Costly

The computer consultant looked around the conference room table at us and answered my question about why he was planning to install a Microsoft Exchange Server for my prospective client. He said, "They have Qwest DSL. Qwest doesn't allow a Port 71 Internet connection. Qwest DSL only works for multiple email accounts by installing an Exchange server."

He added, particularly pleased with himself, "We build the servers ourselves."

When I reported these statements to two highly respected Tucson computer consultants, they had similar reactions. Ed Schaefer, owner of Better Bytes, remarked sarcastically, "I have a few dozen clients who would be shocked to know that their Qwest DSL hasn't actually been working all these years." John Moffatt, a long time technology consultant, sighed and said, "You know, it still amazes me how much bad advice business owners get from computer consultants."

Schaefer and Moffatt said bad computer consultants, driven by ignorance, laziness or greed, do major damage to their clients. Moffatt said, "It's difficult for the average business owner to know their computer consultant's just wrong."

Are you getting bad advice? Here are several red flags.

Your consultant builds PCs for you because, he says, his PCs are better than Dell's or HP's. It's astonishing that otherwise savvy business people believe a guy in Tucson has, for example, Dell's hardware and software buying power, plus Dell's labor costs, quality control and long-term viability.

Schaefer says, "Locally built machines can only compete on price if they're built with inferior parts." Moffatt adds, "One local shop I know of paid for only one Windows license and one Microsoft Office license and then installed that software on all the machines they sold to that business. The customer was pleased with the price but doesn't know the business is illegally using Microsoft and other software, can't get support for it, and takes the risk of getting sued."

Your consultant insists you run Microsoft Exchange server or Microsoft File Server, but you don't understand why. Schaefer has more than 200 clients, the vast majority of whom do not need Exchange server to run their email or File Server to share files. But he has had many clients whose previous consultant sold them those software products and the hardware required to run them.

Moffatt has seen it, too. "Both are fine products, but the consultant has a duty to clients to identify a legitimate business need, and the client should be able to tell you in his or her own words why the technology is needed," Moffatt said.

Your consultant says you need your own Web server. Consultants who try to sell you server and its required services are nearly always driven by greed, not by your needs. Out of 10,000 small businesses, perhaps one in 10,000 might justify its own Web server. The others should let professional Web hosting companies handle the job. Consider this: Even website design and development firms such as NetOutcomes, managing dozens of websites, do not own and run their own Web servers. Again, in large part it's an issue of buying power, not to mention 24/7 tech support.

Your consultant says, "That software just has a bug in it. There's nothing I can do." Moffatt and Schaefer have fixed problems that previous consultants pronounced hopeless. "Sometimes it's just because the guy was too lazy or too embarrassed to call technical support and ask for help," Schaefer said.

Unfortunately, computer consultants still bamboozle people. What would you say if your consultant announced, "You can't get Internet access with dynamic IP addresses. We have to upgrade your account and make Qwest provide a fixed IP address for each machine?" For starters, you can ask for an explanation in plain English. You can also call Qwest and get a second opinion for good measure.

Lawyers, doctors, CPA's, dentists and other professionals must acquire degrees and pass exams. Computer
consultants just need to learn a little lingo and snare a
client or two. So apply an extra dose of common sense.
Check references. Ask a lot of questions. Make sure you
and your consultant have a plan that makes sense.

At the meeting I mentioned above, the company's
trusted computer consultant was looking forward, he
told us, to providing the firm with its own Web server.
He smiled and added that they'd build that machine, too.
The regular business folks around the conference table
smiled uncertainly. One said, "He really takes care of us."
If she'd left out the two words "care of, " then Moffatt,
Schaefer and I could have agreed with her.

Chapter 15

VoIP

Why Choosing a VoIP Phone System Is a Marcom Decision

With marketing and communication (marcom) depending so much on technology, a central question for leaders is this: "How can we draw a line between marcom and technology issues?" The line gets blurred in so many ways that the difference between a marcom and an Information Technology (IT) decision can be hard to determine. One example of the blur is VoIP (Voice Over Internet Protocol) phone systems.

VoIP has become popular for many reasons. A fundamental appeal is simple economics. To use a phone system that is multifeatured, using a traditional carrier-connected landline, an organization usually must make a substantial investment in the phone system itself. VoIP systems, on the other hand, are often sold as "free" as long as the organization agrees to terms involving monthly payments over X number of years and pays separately for the Internet bandwidth required to support VoIP. VoIP systems offer other advantages, such as a greater array of features, than traditional landline phone systems provide.

VoIP systems may also come with some odd-ball issues, including an organization losing its 411 Directory Assistance listing. Depending on the VoIP vendor and how its system is deployed, the organization using VoIP may not be listed in Directory Assistance. A marcom person might think that being listed in Directory Assistance is a no-brainer.

An IT person, on the other hand, might want to argue that with Google and countless other online directory resources available, an organization really does not need to be listed in Directory Assistance. Who is right?

Here's one way to get to an answer. Suppose we were to ask each of four generations this question: "Do you use Directory Assistance?" Or, "Do you ever pick up a phone and dial 411 and ask Directory Assistance for a phone number?" Here are some possible answers, by generation. Gen Y/the Millennials: "Directory what?" (That means "No"). Gen X: "Well, of course I've heard of directory assistance, it but I don't have a landline and my mobile company charges $1 per use, so I never use Directory Assistance." Baby Boomers: "Sure. I still use it from time to time." Silent Generation: "Of course I do. I use it whenever I don't know the phone number."

Perhaps you don't agree with these characterizations by generation. Or perhaps you want to point out some other issue. Great. That just proves the main point here: any discussion about whether a target market uses

Directory Assistance raises a marcom question.

The fact is that organizations that install a VoIP system are not always automatically listed in Directory Assistance. In fact, dig into Google search results and you'll probably find at least one VoIP vendor, call him "Mr. VoIP," who says VoIP makes it impossible to have a Directory Assistance listing. Technically speaking, Mr. VoIP is wrong. The fact is that a capable VoIP vendor can get an organization listed in Directory Assistance. It's also true that VoIP systems can raise other marcom issues, such as voice quality and the handling of incoming calls.

In short, VoIP has a marcom component. So if you're considering the use a VoIP system, of course you want a quality IT person involved, but make sure that marcom has a voice in the decision or you may not like what you hear later.

Mistakes in Directory Assistance Listings Have Big Impact

Heavy traffic makes a business woman think she's going to be late to her appointment at a popular downtown restaurant. She calls 411 Directory Assistance (DA) to get the restaurant's phone number to leave a message, "On the way, running a bit late."

The DA operator says, "I show no listings under that name." A supervisor confirms the "no-listing" results and adds, "Maybe the restaurant went out of business."

A businessman heads off in his car to a meeting but needs to call an out-of-state law firm on the way. He dials DA to get the number. When the DA operator connects him, a fax machine answers and screams in his ear.

What do these two businesses have in common? City and state? No. Type of business? No. Phone companies (carriers)? No.

The restaurateur is perplexed because, when he calls DA from his restaurant, he's given the correct phone number. But some customers have told him his restaurant's not listed in DA.

The law firm gets its phone service through an information technology (IT) consulting firm. The IT company tells the law firm that getting it listed in directory assistance will cost an additional $1,500 and might not work anyway.

The IT company sold the firm its Voice over Internet Protocol (VoIP) phone system. VoIP gives the firm a wide range of handy features without requiring a big, upfront cash expense. The firm has heard a few comments about it not having a DA listing.

An accurate DA listing is still important to have. In 2008, according to one source, people used DA 6 billion times in the United States. As late as 2010, more than one report makes a reference to "billions" of DA calls.

In these two cases, the carrier ("phone company") has screwed up the businesses "phone book" listing. For the restaurant, Cox Communications (the cable company) has made a mistake. For the law firm, the provider of the VoIP service has failed to get the listing completed correctly.

Today, the range of carriers and re-sellers is so wide and deep it's no surprise that mistakes are made. We have dozens, perhaps hundreds, of carriers. Every city has the "incumbent" local exchange carrier (CenturyLink in many Western states) and competitors such as TW Telecom as well as cable companies such as Cox and Comcast. In addition, we have wireless carriers, including AT&T, Verizon Wireless, Sprint, T-Mobile and others who must all provide DA service.

As simple as it might seem to get a listing right, mistakes are made. Some carriers have re-sellers, so the mistake may come from the re-seller. Other carriers just have people on staff who simply don't get a listing entered correctly.

For a business to have its customers get the desired DA experience, carriers complete two steps correctly. First, they must manage their own directory listings and keep them complete and current. Second, they must correctly share that database with all the other carriers.

The repercussions of a bad database being shared extend past DA issues. Websites such as MapQuest and other online directories such as DexPages, YellowBook and WhitePages.com amalgamate data from various sources. These sites then provide phone numbers in their listings. In the law firm's case, that means providing the fax machine number as the voice number.

The solution to the problem begins with double-checking the DA listing by calling DA from more than one carrier — for example, from both a land line and a mobile phone. Next, the organization can just ask its carrier to provide it with a copy of its DA listing, from the database. A screen-shot will do. Then make sure the listing is correct, especially the business name and city and state fields.

When data goes missing or is incorrect, bad things happen. A woman on the way to lunch may have to listen to a Directory Assistance Supervisor suggest the popular restaurant she's headed has gone out of business. Or a man may think he's being connected to a law firm only to have a fax machine answer the phone.

Take a wide variety of carriers, mix in a new technology (VoIP), throw in hundreds of re-sellers, require carriers to share data with each other in specific ways, and the system has just written a recipe that guarantees some bad outcomes. Happily, the restaurateur is so gifted that his restaurant continues to fill up with happy customers, even if Directory Assistance does say he may have gone out of business.

Chapter 16

Website Development

The Two Worst Words to Hear about Your Website

A man I know asked me to look at his company's brand-new website. A bargain hunter, he saw the site as an important sales tool and investment. Our conversation was an opportunity to get free advice. He'd given me some help, a couple of years before, with a commercial office building project. Now it was my turn.

The site seemed visually appealing to his target market, which was especially important for the design/build service he provided. To a search engine, however, his site was a blank slate. Empty. His site was built entirely in Flash.

Search engine robots (bots) crawl through a site to index its content. That content — both the meta tags that only bots see and all the words and images humans see — is used as a significant guide to determining search organic listings and ranking. But at the time his site went live, bots could not index Flash. So the man's brand-new site earned an "F" when it came to search engine optimization.

The Google search we did together turned up all his competitors in the results. Their sites were listed on the first two pages of organic search results, but even in the next several pages his company did not appear at all. In short, the site was a complete disaster. No search performed by a prospective client would produce a listing result. A search using the company's full name did produce one result, but it was completely uninformative.

The fact that his site was invisible to bots was especially bad for this business. His prospective clients were people who had never heard of his company before.

Shaken, he asked me what the best solution would be. I thought back to a meeting I had included him in a few years earlier. We met to evaluate a residential home for sale on a major artery. The question was whether the house could be converted to an office building. I could not imagine how the house could be converted, but construction is not my area or expertise. That's why I had experts with me. He was one of them. He stood on the site thinking about building code requirements and construction options.

Compared with real-world design and construction, all organizations face far more uncertainty in making decisions about website development. In website development, contractor's licenses do not exist. No plans must be approved by a government entity. No zoning or building codes apply. No inspectors visit the site to examine the plans and approve the work.

The very next day after looking at this all-Flash, zero-results site, I got an invitation to attend a conference for principals, partners and presidents of ad agencies. The three-day conference's first bullet point caught my eye. The session topic was "Selling Stuff You've Never Done Before but You Need to Be Selling."

The invitation made me revisit the design/build man's all-Flash website. Clearly the website developer had not attended a conference of similar quality. Amazingly, the site he'd sold to the design/build man was a clone of his own site. The code was identical. Only the images had been changed.

Knowledgeable readers may already have been thinking, "Nobody does an entire website in Flash anymore, so why complain about it?" It's just an example. We've seen complete failures in competence with sites written in ASP, PHP or other languages.

This man, an expert in real-world construction, wanted to know what could be "done about" his website being invisible. I thought back to the advice he'd given me years ago when we looked at that useless house on that great piece of commercial real estate.

At the time, he stood here, made a sweeping gesture with his hand and said, "Scrape it." In construction, "scrape it" means wipe it off the face of the earth and start over again, on an empty lot.

Now he was asking me, "What do I do about my website?" I gave him the same answer. "Scrape it."

He was shaken, speechless. Then he said, "Really?"

I added, "Well, you can still use the look. But the site has to be rebuilt from the ground up."

What can and should you do in the planning stage to prevent having to scrape your own website? Common sense applies. Get educated. Learn the language and as much as you can about website development standards. Get more than one proposal. Ask questions. Require answers in plain English. Check references. Get site specifications written out. Find a way to understand them. Trust your gut instincts.

No government agency is going to be checking out the site built for you. The due diligence, every step of the way, is up to you.

Did You Order SEO, Google Analytics and Facebook ROI on the Side?

The owners of a brand new, custom built Tucson home were upset. In the first few days after they'd moved in, they'd spotted seven scorpions scampering around.

"That's nothing," a client of mine from years ago said. "In one new home, I killed 12 scorpions in one evening." He explained that homebuilders leave new homes open to insects and rodents right up until the last few weeks of construction. When doors and windows are finally closed, insects get trapped inside. Worse yet, new homes are routinely left with gaping routes for pests to enter. The key entry points include a gap between the frame and stucco wall and the foundation, door thresholds, can lights over porches and rain weep holes on windows.

Similarly, the launch of a new website may also leave "holes" and produce unfortunate surprises. A while back a marketing director we know oversaw the launch of a gorgeous, brand-new website and then immediately demanded to know why certain services had not been included. True, the cheap web-hosting service that the client insisted on did provide visitor reports, but why wasn't Google Analytics automatically included? And why wasn't great Search Engine Optimization (SEO) included, SEO so powerful that the new site would show up on Google from Day 1 of Launch with any key-word search?

The key to getting what you expect out of a new website, a remodel of an existing site, an email campaign, an SEO project or a social media campaign is to get educated and then make sure the proposal you sign off on specifies exactly what you're going to get and why. If the developer's proposal doesn't include SEO, and you want SEO, then have a discussion and decide on what level of SEO offers you the best Return on Investment (ROI). It's actually possible to spend more on SEO than on the website that mere mortals — human beings, that is — are looking at.

The same approach should be taken if you're expecting a Facebook campaign to produce results for you. Identify the results you want, the metrics you're going to use to measure those results and the project work that's going to produce that outcome.

Email campaigns — which can be highly effective, by the way — should also have detailed specifications. Key email campaign considerations include list acquisition and development; campaign management and growth; and opens, unsubscribes, forwards and click-throughs. Your plan should detail who is handling your strategy, managing the ongoing campaign, producing content and reporting results. When in doubt, ask questions. If you don't understand the answers you're getting, talk to other vendors and keep asking until you get a clear picture. Educate yourself through online research. Of course the process can be time-consuming, but consider how much time it takes to make all the decisions involved in buying and moving into a new home.

That client of mine who was in the pest control business was carving out a niche back then in "exclusion work." The benefit of exclusion work is simple: seal up a home properly and you drastically reduce the number of times that an exterminator will have to treat the interior.

At that time, we initially saw homebuilders as possible buyers of exclusion work. The benefit to the homebuilder, we thought, would be that new homeowners would not be shocked by scorpions scampering around or other kinds of pest problems arising immediately. Sadly, not one homebuilder who was contacted had an interest in offering exclusion work as an upgrade. The homebuilders seemed to feel that by suggesting exclusion work as an add-on expense, the builder was admitting the home wasn't 100 percent sealed up in the first place. On the other hand, homebuilders who pay for exclusion work add a cost without getting a benefit that helps sell the home.

Website builders may be guilty of looking at SEO, Google Analytics and other add-ons the same way. Offering a menu of options as add-ons and prospects may demand that most or all of the upgrades be included for the same price.

In the end, the age-old phrase caveat emptor — let the buyer beware — applies even more to buying a new website than to buying a new home. In homebuilding, the buyer is partly protected by city, county, state and federal codes or laws. There's no such protection when buying a new website, so buyers must accept responsibility for making sure they know what they're paying for. For an organization, the consequences of not being vigilant about website development and other Internet marketing efforts can be much more severe than the risk of getting stung by scorpions.

Reading Glasses and the Ever-Changing Width of Websites

Years ago a member of the website committee for a large client of ours complained about having to scroll horizontally to view the company's site. She was a great client, but low-tech, so I visited her office and confirmed that the resolution on her monitor was set extremely low. As a result, the site on her monitor looked huge, thus the horrible need to scroll left and right just to see the whole page. The text on the site was huge. I switched her settings to standard resolution and showed her how the site now fit nicely on the screen.

"Change it back," she commanded. "I don't want to have to wear my glasses to read it."

Today, people still have monitors — laptop or desktop — set at a wide range of resolutions, most ranging from 1,280 to 1,920 pixels wide. Screen resolution matters because one goal of website development is to make optimal use of the available real estate. That real estate is measured in pixels.

After the first few years of site development, website programmers settled on a pixel width of 645. Why 645? That (now) narrow width meant that a website page, if printed on a printer with a "portrait" setting, would fit on 8.5-by-11-inch paper. Most people were absolutely sure their site visitors printed pages.

Years later, two factors changed the ideal pixel width. One change was that some code that could now be used converted site pages to a "printer-friendly" format. A popular approach to "printer-friendly" pages was to omit the navigation. There is no need to see the navigation on a hard copy, after all. This omission created more space for the content of each page. The code guaranteed that with proper planning and the right code, printing was always successful.

Another major factor that changed ideal pixel width was monitor size. Monitors not only got bigger and bigger but also became much more horizontal. Today, some people even use HDTVs of various sizes to view sites.

All these developments, taken together, have produced a new standard site width. The main content of a site (excluding background images) typically has a width of 960 to 998 pixels. Even at 998, if your monitor is set to the "normal" resolution of 1,280, the new "standard-width" sites will not quite take up all of your monitor's available space. Some computer users may even have the resolution set at 1,920, which means that most websites will have a lot of empty space around them.

With the proliferation of screen sizes out there –

ranging from a few inches on a mobile phone to 20 inch-es or more on a desktop monitor – some developers have started to use what has been dubbed "responsive" design. Here are two approaches to responsive design.

Special, complex programming can turn a website into a "chameleon" in size, changing the site's width ac-cording to the specific screen-resolution setting of the machine being used. Google News and Amazon are two examples of sites that scale automatically. With this ap-proach, however, design options are limited, the effort is complex and the price tag is substantial. For the vast majority of organizations, this variable-width approach is not an option. On the plus side, affordable website pro-gramming can at least recognize whether the site is being delivered to a smartphone or a computer and then deliver a site that accommodates the device. A "smartphone site" has to be developed as a separate design-and-build proj-ect, usually with a link to the "full site" as an alternative.

Apple made this "website width issue" slightly more complicated when it began using its Retina display tech-nology to produce an extremely high resolution on the iPad. Then Apple's newest MacBook Pro used the same Retina display screen technology. The screen resolution is a stunning 2,880-by-1,800-pixels. For technical rea-sons, that 2,880 doesn't ruin our width standard of 960 to 990 pixel . Even so, it's easy to see that this is what's com-ing. Higher-resolution monitors are inevitable. Some day the current standard width will give way to wider sites.

Meanwhile, any site not set for today's standard will

stick out like a sore thumb. One way to measure your own website's width — or the competition's — is to get a pixel ruler online. Then open the ruler on the site you care about and see how the site measures up. If your site doesn't, maybe it's time for a change.

Here's more good news, for that former client of ours who didn't want to have to put on her reading glasses. Today, Windows 7 allows users to increase the font size on the screen, to 125 percent or even 150 percent of actual size. That means she'll be able to view sites without her reading glasses, and without scrolling from side to side.

When Cousin Billy Builds Your Website

Budget-conscious companies often turn to a friend or relative to build their website for free. Someone's daughter, future son-in-law, cousin or personal friend wants to learn how to build websites and volunteers for the job.

The problem is that the Cousin Billy Website is more likely to be a disaster than a success. This disaster makes the company look bad.

If you plan to have Cousin Billy build your organization's website, improve your chances of success by following these guidelines.

1. Pay Billy $1,000 or more.

Why pay Billy when he'll do the job for free? For starters, if you doubt that Billy's work is going to be worth $1,000, then why are you wasting your time and his on the project? Second, if $1,000 seems to you like a lot of money for a new site, your commitment to a new site's success is clearly too low. Don't bother getting started.

Another reason to pay Billy is to get the revisions you need. It's hard to hold a guy's feet to the fire when he is, after all, working for free. If you pay him, you can have a clear conscience about making him do the work right.

Lastly, pay him in fourths. One fourth up front, one fourth to see the design, one fourth to finish the site and allow you to test it, and the last fourth to launch the site (which means more testing).

2. Make sure you choose a professional hosting service.

Cousin Billy may think he's doing you a favor by getting you hosting from a "Brand X" company for $2.95 per month. Typically that kind of site performs badly and/or the tech support is terrible.

Do a little research of your own. Pick a top-rated hosting firm.

3. Know how to access your own site, and keep a set of site files on your computer.

Surprising numbers of businesses we talk to don't know where their site is being hosted, how to access their website files, or where their back-up files are. Make sure you know.

4. Choose your own domain name.

Billy may know HTML, but that doesn't make him a marketing expert. He is almost certainly not qualified to provide you with the best website address. We've seen some pretty dumb domain names selected by Cousin Billy and accepted by lazy or ignorant companies. Get domain name advice from a marketing and communication professional.

5. Make sure you know how long Billy's going to be around.

If you have questions about the site or need changes a month or a year later, how will you reach Billy? What if he's no longer your daughter's boyfriend? Then what will you do?

6. Judge Billy by the same standards you use for other company work.

Just because Billy's working for $10 an hour, don't settle for bad work. If the purpose of your Website is to get more customers, then get half a dozen of your current customers to look at the site and tell you how they like it. Their opinion is more important than your own.

Remember, your site represents your organization and creates an image and market position for you, in your city, state, the United States and around the world.

7. Plan ahead.

Once Cousin Billy launches your site, who is going to keep it current? Launching a website is like getting a dog — it's initially exciting, but dogs, and websites, need routine care. In some ways, dogs are easier to take care of — they sleep most of the time. Sites are up 24-hours-a-day. Decide who will revise the site on a regular basis. Understand the skill set and software needed for site maintenance.

8. Test the site with popular web browsers like Internet Explorer, Firefox, Chrome and Safari.

We know of one company grossing over $1 million a month that thought its Cousin Billy site was just fine. Cousin Billy was looking at the site only with Internet Explorer. Everyone at the company was, too. As it happened, the site had major problems when viewed with the other major browser. The result: millions and millions of people weren't able to use the site well.

9. Ask Cousin Billy what he's going to do about Search Engine Optimization (SEO).

If he says he wants you to do that, then at least you know to budget for SEO as an additional expense. If your site can't be found, it won't be seen. Get expert advice on SEO.

10. Tell Cousin Billy to avoid bells and whistles.

The Cousin Billys of the world are prone to using gadgets that are in bad taste or dysfunctional.

11. Have Billy run the W3C Code Validation Service on your new site.

The World Wide Web Consortium (W3C) is an international consortium founded in 1994 to develop site guidelines and specifications and do other important work. Its Code Validation Service (CVS) is free and is a valuable measurement tool for quality programming.

If your site has 50-plus errors per page (we've seen sites with 300-pluserrors), then each page needs to be cleaned up. Billy can and should run the CVS himself, of course, but if you're taking the Cousin Billy Website route, then you'd better run the CVS yourself, or have an employee do so and report results.

Follow these guidelines to increase your chances of getting a site with some actual value for your investment. Better yet, use these guidelines to write up an agreement that you and Cousin Billy sign before he gets his first $250 check. You'll be glad you did.

Chapter 17

Website Hosting

Website Hosting: When Ignorance Is Not Bliss

At a two-day national conference on digital marketing, one speaker's topic focused on ways a firm could keep its clients by tying them up. Among his recommendations: own and control your client's web hosting.

Countless organizations we've talked to are afraid that choosing us to take over their online marketing will cause their current firm to punish them in some way. Maybe their site will go dark. Maybe files will be lost. Maybe their email will quit working. They are just not in control. The vendor is.

An organization can lose control of its own website hosting even when the site is hosted internally. Here's an example. The head of an internationally-known nonprofit organization told me one day, "Our email went down because our website got hacked." He wanted advice, so we did some quick research.

To save $100, his organization was hosting its web site on its own Microsoft Windows Small Business Server

(SBS). SBS can be a great server for an internal network. Adding "web site hosting" to its list of services provided, however, can cause problems.

We urged the exec of the international nonprofit to have the site moved, immediately, to a national hosting firm. The $100 per year cost was small compared to the added value of security — one security lapse had already occurred.

He said, "I talked to my IT guy (an employee) and he's got it fixed."

Executives accept these two bad choices — external control of site hosting and/or very poor internal control of hosting — for many reasons. One reason is ignorance. Leaders seem to abdicate decision-making because the terminology, like "hosting," "server" or "bandwidth," intimidates them. The truth is that website hosting is easy and vital for owners, leaders and managers to understand.

Websites live on computers connected to the Internet. The computers host a website and "serve it up" when anyone uses the Internet and types in the domain name — that address — of that website. Hosting comes in a wide variety of types, including self-managed rack space, managed and dedicated, VPS and shared.

Perhaps more than 90 percent of all small organizations use "shared" hosting. "Shared" means that websites share space and resources of a single computer, a computer that may easily handle hosting a thousand or more sites.

Today what still surprises us is the number or organizations that let their IT people talk them into "saving money" by hosting their site on their own Small Business Server. The organization "saves" $90 a year and gives up control to the person/vendor who is running the Small Business Server.

The cost of abdication is high. For starters, the same computer (server) that contains an organization's vital computer files is the computer that hackers find most tempting to attack. Succeed and a hacker may get access to an entire internal network.

Two, if the network server goes down, the organization's website goes down with it, a double crisis. Three, if an organization's Internet connection (nearly all small businesses pay just one ISP) fails, the organization's website and email access fails. That's another double crisis, which makes for a very dark day. Four, the time it takes to program and support a web server is certainly not free, no matter what the vendor claims. Five, an organization that hosts its own website must pay for a separate solution for email spam and virus protection.

Some computer network consultants like the idea of controlling a client's site by placing it on the Small Business Server they are already managing. Ad agencies and site developers who want to control clients, on the other hand, are more likely to use a Virtual Private Server (VPS) to host all their clients' sites. VPS packages cost as little as $40 a month and provide enough space and services to host as many as 100 to 200 websites. That's cheap control.

The majority of small organizations find shared hosting is the best choice. That's why organizations should own their own web hosting, trust in a national firm and provide their website developer with access to the site.

Without professional hosting and direct ownership, bad things can happen. One organization came to us in an emergency. Its site had gone offline because the developer hadn't paid the hosting bill! At another organization, its website files had been taken hostage until it paid a substantial "file release licensing fee."

Now, remember the organization whose top executive said his IT guy had the hacking issue resolved? Sadly, two days later that site was successfully hacked again, only this time the hacker didn't stop with deleting website files. Thousands of other vital files vanished from the server.

With potential damage so high, and the cost of professional hosting so low, top management should take the time to learn the basics about web hosting and even get a second or even a third opinion. Take control or be controlled. That's the choice that's being served up to you.

Chapter 18

Website Programming

The Wrong Website Language Can Squeeze You to Death

The bad news about the organization's website stunned the marketing director. The site was essential to serving current customers and attracting new ones, but the company's website programmer could not make vital changes to the site.

The programmer explained, "The only guy who knew how to do that left." Top management couldn't find anyone, even outside of the organization, who could make the much-needed changes.

Weeks later, we were collaborating with a different organization. The marketing director said, "We can't make those changes on our site until a couple of weeks from now."

I said, "Why two weeks?"

Here's what we learned. In the organization's IT department of more than 40 people, only one person could write the code required to edit the site. Unfortunately, that guy was seriously ill, had already missed weeks of work and wasn't expected back for a while.

These two organizations shared the same problem. Both had websites written in a programming language so obscure that no programmer could be found to do the work.

Leaders and managers in all kinds of organizations must deal with a key web site development question. How can you be sure your site is built using popular, common, mainstream code, code that most site programmers know? How can you be sure, if the original developer is ill or leaves, someone else can manage the site?

For starters, make sure you do not have a truly compelling reason to make a nonstandard choice. If not, then make an obvious choice. Choose between the Coke and Pepsi of website programming: PHP (Coke) and ASP. NET (Pepsi).

Aside from Fortune 1000 companies, the majority of websites used by small organizations today are built with one of two languages. The two mainstream choices are PHP (Hypertext Preprocessor) or ASP.NET (Active Server Pages).

If you're thinking, "But websites are written in HTML (HyperText Markup Language)," you're right. PHP and ASP.NET generate HTML.

These two languages are popular, mainstream choices for many reasons. For one, the talent pool of experienced programmers is relatively large. If your developer quits or gets ill or is terminated, you'll find it easier to find a suitable replacement.

Another advantage is the fact that widespread use of these two programs creates rich resources programmers can use. These include documentation (how-to) and sample scripts (bits and pieces of website programming commonly used).

Third, the two most popular types of hosting — Linux and Windows — easily support PHP (Linux) and ASP.NET (Windows).

Another factor in choosing a programming language is whether and what kind of database the site must use. Organizations with an MS SQL (Microsoft Structured Query Language) database in place often choose ASP.NET. PHP can work well with MS SQL, but PHP is an even a better choice for MySQL databases.

The next challenging decision an organization makes is the choice of a programmer. Choosing a programmer is genuinely similar to choosing a writer to write wording about your organization or choosing an accountant to manage the books.

A writer can choose to use Word, WordPerfect or even Apple's iWork as the word processing program, but this choice has nothing to do with the quality of the writing. Similarly, the fact that a person knows how to use QuickBooks does not mean that person will be a good full-charge bookkeeper.

These programs are just toolboxes full of tools. For a writer, a bookkeeper and a site programmer several vital factors determine the quality of the product: skill, talent, effort and experience.

Some aspects of programming involve skills that can be learned. Talent is also a major factor. Some programmers are more talented than others. Some programmers write slowly, some quickly, some awkwardly, some gracefully. Some are mistake-prone, others error-free.

How can a nonexpert manager evaluate the skill and talent of a programmer? It isn't easy, but the same approach used to select people for other positions should work.

Here's one useful step to try. Use the W3C Code Validator at *http://validator.w3.org/* to check the quality of a programmer's code. Nearly all pages may have some errors, but the home page for the City of Phoenix, *www.phoenix.gov*, when we checked, had 214 Errors and 2 warnings. That's bad.

You should also determine whether the programmer wants to use a language based on what you need or what he or she likes to use. Lastly, get a second opinion, and even a third opinion.

One of the marketing directors I mentioned earlier said about his organization and its bad programming choice, "I see why it's called Python. It squeezes you until you're dead." Staying alive on the job is always a good idea, so take the time to make the right choice about the programming language that runs your site.

Afterword

In the preface of her dazzling book, *Slouching Towards Bethlehem*, Joan Didion tells her readers to keep in mind that, "writers are always selling somebody out." I've remembered that line ever since I first read the book about 30 years ago. It rings true, now, about me and this book.

Often, in this book I'm selling myself out. Sometimes I'm doing that by revealing more about myself, personally, than my wife, a very private person, is comfortable having me share. To her I apologize and ask for forgiveness. Without her, so much else I've written, and this book, wouldn't be possible.

Sometimes in this book I'm selling myself out professionally. By writing about an aspect of people and technology and/or marketing, I admit I was struggling to understand the topic in the first place.

In selecting each of these pieces of writing, my thinking was that if I had to struggle to understand a problem or paradox, then probably lots of other people did, too.

Sometimes I know I've connected with readers. One column caused a marketing director to email me this comment, "You are one funny guy. I just read your column . . . What a classic! It's hilarious, it brought tears to my eyes, made my belly hurt and blew my concentration for the whole rest of the day!"

Thankfully, none of the prospective clients or clients involved in these topics has ever commented or complained. If I have been guilty of selling them out, it seems they either don't know or don't care.

This absence of complaint has caused me, from time to time, to wonder whether anyone's reading the newspaper or the blog or the e-newsletter. Then I get reactions, and some affirmations.

A man in a grocery store accosted me once as I selected oranges and told me, "I read every column you write, but I almost never agree with you."

In a Starbucks, a breakfast buffet line, or elsewhere, people comment. "Aren't you Dave Tedlock? I remember that column you wrote on ..."

As I put this book together I set aside dozens and dozens of columns that, in retrospect, seemed dated or feeble in one way or another. I wondered why I'd written them in the first place and why my editors had happily placed them in a newspaper or magazine.

Every column included in this book, on the other hand, earned some positive feedback. The one about taking a vacation with or without technology, for example, prompted three people to immediately email me gracious praise and several other people to thank me in person.

My favorite comment of all time came from a government worker, a mid-level manager, who wrote me she loved "every word. You rock." Her comment alone motivated me to spend some extra hours making this book worthwhile.

Sometimes, when I worry about "selling somebody out" in a column, I remind myself of my experience with our long-time Santa Fe friend and neighbor, Archie West. Archie's a life-long cattle rancher and genuine cowboy, a whole book unto himself. One day he visited us in our Santa Fe home and our conversation led to the piece, "Living Our Lives Online, Not Written in Stone."

Archie does not read business newspapers from Tucson, or anywhere else, so I decided to read that piece to him, right then. He seemed happy to listen, so I looked at the copy on my laptop and read it out loud while he and my son and our dog, Jazz, listened. As I read, Archie's face broke into a big smile. Then he beamed and laughed out loud exactly the same way he had when my son wrote his name in concrete and my wife took a picture of Jazz's impromptu signature as the evening light faded.

Dave Tedlock is the head of NetOutcomes, a digital marketing firm. For years he has written a marketing and technology column for various publications, including the Tucson Citizen's Tucson Business Edge, Idaho Business Journal, Inside Tucson Business, and The New Mexico Business Weekly.

Tedlock taught writing and business communication for eight years in many universities, including the Harvard Business School and Iowa State University. For 13 years he worked in ad agencies as a copywriter, account manager or creative director.

Tedlock has published short stories, scholarly articles and a writing textbook (with Paul Jarvie). He earned a Master's degree in Fiction Writing from Brown University. He lives in Tucson and Santa Fe.